ENCYCLOPAEDIA OF OSCILLATIONS AND WAVES

ENCYCLOPAEDIA OF OSCILLATIONS AND WAVES

Vol. 2

Manpal Singh

ANMOL PUBLICATIONS PVT. LTD.
NEW DELHI - 110 002 (INDIA)

ANMOL PUBLICATIONS PVT. LTD.
H.O.: 4374/4B, Ansari Road, Darya Ganj,
New Delhi-110 002 (India)
Ph.: 23278000, 23261597
B.O.: No. 1015, Ist Main Road, BSK IIIrd Stage
IIIrd Phase, IIIrd Block,
Bangalore - 560 085 (India)
Visit us at: www.anmolpublications.com

Encyclopaedia of Oscillations and Waves

First Published, 2008

ISBN 978-81-261-3541-7 (Set)

PRINTED IN INDIA

Printed at Mehra Offset Press, Delhi.

Contents

Preface

Simple Harmonic Oscillation is defined as the oscillation of a moving body, which is repeated after a definite interval of time. In oscillation, the moving body has the maximum velocity at the centre point and minimum velocity on the end points of the moving path. The moving bodies have velocity, acceleration, displacement amplitude and periodic time. Acceleration of the body is defined as the change in velocity per unit time. Displacement is defined as the distance of the moving body at any time from the centre point of the moving path. Amplitude of the moving body is defined as the maximum displacement of that body, on one side of the moving path, from its centre point. The acceleration is directly proportional to the displacement, but always directed towards the centre point. The time taken by the moving body in one complete oscillation is defined as the periodic time of the body. The periodic time depends upon effective length of the thread, from which the oscillatory body is hanged and on gravitational acceleration. In case of spring motion, the periodic time depends on the constant force of the spring and mass of the body.

When an object is set into oscillations, without any external influence, its amplitude of oscillation gradually decreases to zero. This is because the oscillating object is always subject to frictional force, arising from air resistance or from within the object itself. The force applied against motion is dissipated in form of heat out of the object only. When an oscillation body is displaced from its equilibrium-position and released, it begins to oscillate. The frequency of oscillations depends upon certain properties of the body. Such oscillations are called free oscillations and their frequency is called 'natural frequency' of the body. When a body is set into oscillation by an external periodic force of the same frequency as the natural frequency of the body, the amplitude of oscillation is increased considerably. This phenomena is called resonance. Hence the resonance is a special process of forced oscillation.

When the frequency of the external force is equal to the natural frequency of the body, then at each step the force helps the body in going ahead in its oscillation. The successive impulse given by the periodic force is added up and increase the amplitude of oscillation of the body continuously. The loss of energy of the body is due to air resistance and internal friction and it also increases with the increase of its amplitude of oscillation. When two harmonic wave-trains of the same frequency, having a constant phase relation travel simultaneously through a medium, the resultant wave intensity at any point is different from the sum of the intensities caused by separate wave-trains. This is the phenomenon of interference. When two harmonic wave-trains of the same frequency travel along the same line in opposite directions, their super position gives a pattern having alternately nodal points of zero displacement along with additional points of maximum displacement. This is known as 'stationary-wave pattern'. A practical source of light has a huge number of atoms, which emit discontinuously the trains of light waves at short intervals. These intervals are different for different atoms. Therefore, the phase of the light wave-train emitted by the innumerable atoms in the source changes with time in a random manner.

This exclusive work has a perfect account of harmonic oscillations in an interesting manner. Let's hope, this work would be able to serve its purpose by being beneficial for scientists, professionals, students and of course the general readers.

Editor

1

Introduction

Oscillation is the variation, typically in time, of some measure about a central value (often a point of equilibrium) or between two or more different states. Familiar examples include a swinging pendulum and AC power. The term vibration is sometimes used more narrowly to mean a mechanical oscillation but sometimes is used to be synonymous with "oscillation". Oscillations occur not only in physical systems but also in biological systems and in human society.

Simple Systems

The simplest mechanical oscillating system is a mass attached to a linear spring, subject to no other forces; except for the point of equilibrium, this system is equivalent to the same one subject to a constant force such as gravity. Such a system may be approximated on an air table or ice surface. The system is in an equilibrium state when the spring is unstretched. If the system is displaced from the equilibrium, there is a net *restoring force* on the mass, tending to bring it back to equilibrium. However, in moving the mass back to the equilibrium position, it has acquired momentum which keeps it moving beyond that position, establishing a new restoring force in the opposite sense. The time taken for an oscillation to occur is often referred to as the oscillatory period.

The specific dynamics of this spring-mass system are described mathematically by the simple harmonic oscillator and the regular periodic motion is known as *simple harmonic motion*. In the spring-mass system, oscillations occur because, at the static equilibrium displacement, the mass has kinetic energy which is converted into potential energy stored in the spring at the extremes of its path. The spring-mass system illustrates some common features of oscillation, namely the existence of an equilibrium and the presence of a restoring force which grows stronger the further the system deviates from equilibrium.

The harmonic oscillator offers a model of many more complicated types of oscillation and can be extended by the use of Fourier analysis.

Damped, Driven and Self-induced Oscillations

In real-world systems, the second law of thermodynamics dictates that there is some continual and inevitable conversion of energy into the thermal energy of the environment. Thus, *damped* oscillations tend to decay with time unless there is some net source of energy in the system. The simplest description of this decay process can be illustrated by the harmonic oscillator. In addition, an oscillating system may be subject to some external force (often sinusoidal), as when an AC circuit is connected to an outside power source. In this case the oscillation is said to be driven.

Some systems can be excited by energy transfer from the environment. This transfer typically occurs where systems are embedded in some fluid flow. For example, the phenomenon of flutter in aerodynamics occurs when an arbitrarily small displacement of an aircraft wing (from its equilibrium) results in an increase in the angle of attack of the wing on the air flow and a consequential increase in lift coefficient, leading to a still greater displacement. At sufficiently large displacements, the stiffness of the wing dominates to provide the restoring force that enables an oscillation.

Coupled Oscillations

The harmonic oscillator and the systems it models have a single degree of freedom. More complicated systems have more degrees of freedom, for example two masses and three springs (each mass being attached to fixed points and to each other). In such cases, the behaviour of each variable influences that of the others. This leads to a *coupling* of the oscillations of the individual degrees of freedom. For example, two pendulum clocks mounted on a common wall will tend to synchronise. The apparent motions of the individual oscillations typically appears very complicated but a more economic, computationally simpler and conceptually deeper description is given by resolving the motion into normal modes.

Continuous Systems-waves

As the number of degrees of freedom becomes arbitrarily large, a system approaches continuity; examples include a string or the surface of a body of water. Such systems have (in the classical limit) an infinite number of normal modes and their oscillations occur in the form of waves that can characteristically propagate.

2

Simple Harmonic Oscillations

Any motion which is repeated after equal intervals of time is called 'periodic motion' or 'harmonic motion'. If a particle having periodic motion moves back and forth over the same path, it is said to be 'oscillating' or 'vibrating'. The pendulum of a clock swings back and forth and is said to perform an oscillatory motion. Similarly, the motion of the prongs of a tuning fork, the internal motion of atoms within molecules, the motion of electric and magnetic fields in an electromagnetic wave are the examples of oscillatory motion.

Periodic Time: The 'periodic time' T of an oscillatory motion is defined as the time taken for one complete round trip of the motion, *i.e.* for one oscillation (or one cycle).

Frequency: The frequency of the motion, n, is defined as the number of oscillations in 1 second. Obviously, the frequency is the reciprocal of the periodic time.

$$n = \frac{1}{T} \text{ (\textit{cycles per} sec.)}$$

Displacement: There is a position for every oscillating particle at which no net force acts on the particle. This is called the 'equilibrium position' of the particle. The distance of the particle in any direction from its equilibrium position at any instant is called the 'displacement' of the particle in that direction at that instant. The displacement changes periodically on magnitude and direction.

Amplitude: The 'amplitude' a of the oscillations is the distance between the equilibrium position and an extreme position. It is the maximum displacement.

Phase: The 'phase' of an oscillating particle at any instant defines the stage of the particle as regards its position and direction of motion at that instant.

Mechanical Energy

The total mechanical energy E of an oscillating particle consists of kinetic energy K and potential energy U.

$$E = K + U.$$

The particle has kinetic energy by virtue of its motion, and potential energy by virtue of its displacement from the equilibrium position.

If there are no non-conservative forces (like friction) acting on the particle, the total energy E remains constant.

Restoring Force: The oscillating particle can remain at rest in its equilibrium position at which no net force acts upon it. When it is displaced from this position, a periodic force acts upon it in such a direction as to return it to its equilibrium position. This is called the 'restoring force' F.

We know that the potential energy can also be defined as a function of position whose negative gradient gives the intrinsic force. Hence the (restoring) force acting on the particle at any position is derivable from the potential energy function:

$$F = -\frac{du}{dx}$$

$$x = C_1 (\cos \omega t + j \sin \omega t) + C_2 (\cos \omega t - j \sin \omega t)$$
$$= (C_1 + C_2) \cos \omega t + j (C_1 - C_2) \sin \omega t.$$

Let us make a change in arbitrary constants by putting

$$C_1 + C_2 = a \sin \phi$$

and $$j (C_1 - C_2) = a \cos \phi$$

where a and ϕ are new constants. This gives

$$x = a \sin\phi \cos \omega t + a \cos\phi \sin \omega t$$

or $$\boxed{x = a \sin(\omega t + \phi).} \qquad ...(ii)$$

This is, in fact, a solution of the equation of S.H.M. The values of the constants a and ϕ depend upon how the motion is started,

Amplitude and Phase: In the above equation the term sin $(\omega t + \phi)$ takes on values from +1 to – 1. Hence the maximum value of the displacement x is a which is the "amplitude" of the motion.

The quantity $(\omega t + \phi)$ is called the "phase" of the motion. The constant ϕ is called the "phase constant".

Velocity: The velocity u of the oscillating particle can be obtained by differentiating eq. (ii). Thus

$$u = \frac{dx}{dt} = \omega a \cos (\omega t + \phi) = \omega \sqrt{(a^2 - x^2)}.$$

This shows that the velocity is maximum $(= \omega a)$ when the displacement x (and hence the force F) is zero and vice-versa.

Periodic Time: If the time t in eq. (ii) is increased by $2\pi/\omega$, the displacement becomes

$$x = a \sin \left[\omega(t + 2\pi/\omega) + \phi\right]$$
$$= a \sin (\omega t + 2\pi + \phi)$$
$$= a \sin (\omega t + \phi).$$

That is, the displacement repeats itself after a time $2\pi/\omega$. There-fore, $2\pi/\omega$. is the "period" of the motion, *T*. Thus

$$T=\frac{2\pi}{\omega}.$$

But $\omega^2=\frac{k}{m}$. Therefore the periodic time is

$$T=2\pi\sqrt{\left(\frac{m}{k}\right)} \qquad \text{(iii)}$$

$$=2\pi\sqrt{\left(\frac{\text{mass}}{\text{force per uint displacenent}}\right)}$$

$$=2\pi\sqrt{\left(\frac{1}{\text{acceleration per uint displacenent}}\right)}$$

$$=2\pi\sqrt{\left(\frac{\text{displacenent}}{\text{acceleration}}\right)}.$$

The eq. (iii) shows that the period is independent of the amplitude of the motion.

Frequency: The frequency *n* of the oscillator is

$$n=\frac{1}{T}=\frac{\omega}{2\pi}=\frac{1}{2\pi}\sqrt{\left(\frac{k}{m}\right)}.$$

$\omega\,(=2\pi n)$ is called the angular frequency.

Relations between Displacement, Velocity and Acceleration

The displacement of a particle in S. H. M. is given by eq. (ii) *i.e.*

$$x=a\sin(\omega t+\phi).$$

Let us take, for simplicity, $\phi=0$ and put $\omega=\frac{2\pi}{T}$. Then

$$x = a \sin \frac{2\pi t}{T}$$

Thus at	$t=$	0	$T/4$	$T/2$	$3T/4$	T
we have	$X=$	0	a	0	$-a$	0

Plotting x against t, we obtain a "displacement curve" as shown in figure given below

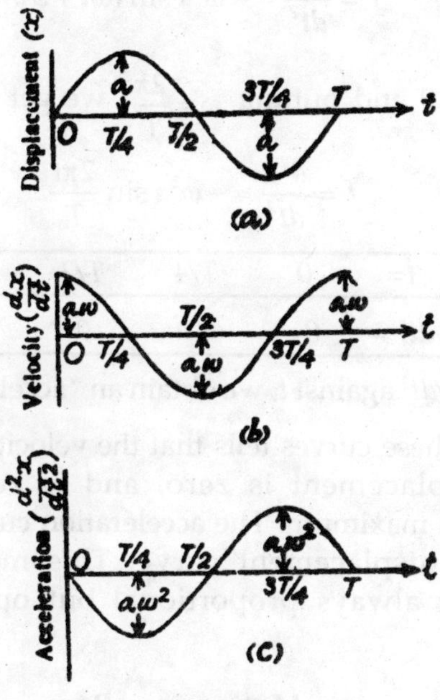

(Fig.)

Now, the velocity of a particle in S. H. M. is given by

$$u = \frac{dx}{dt} = \omega a \cos(\omega t + \phi)$$

Again, taking $\phi = 0$ and putting $\omega = \dfrac{2\pi}{T}$, we get

$$u = \frac{dx}{dt} = \omega a \cos \frac{2\pi t}{T}$$

Thus at $t=$	0	$T/4$	$T/2$	$3T/4$	T
we have $dx/dt =$	ωa	0	$-\omega a$	0	ωa

Plotting dx/dt against f, we obtain a "velocity curve".

Again, the acceleration of a particle in S. H. M. is obtained by differentiating eq. (ii) twice, *i.e.*

$$f = \frac{d^2x}{dt^2} = -\omega^2 a \sin(\omega t + \phi).$$

Taking $\phi = 0$ and putting $\omega = \frac{2\pi}{T}$, we get

$$f = \frac{d^2x}{dt^2} = -\omega^2 a \sin\frac{2\pi t}{T}.$$

Thus at $t=$	0	$T/4$	$T/2$	$3T/4$	T
we have $d^2x/dt^2 =$	0	$-\omega^2 a$	0	$\omega^2 a$	0

Plotting d^2x/dt^2 against t, we obtain an "acceleration curve".

A look on these curves tells that the velocity is maximum when the displacement is zero, and is zero when the displacement is maximum. The acceleration curve is a mirror image of the displacement curve. This means that the acceleration is always proportional but opposite to the displacement.

Exact Simple Harmonic Motions are Rare

The simple harmonic motion is characterised by the 'linear' relationship between the restoring force and the displacement. In actual systems this relationship is not exactly linear, and the systems can be treated as simple harmonic oscillators only *within certain limits*. For example, the motion of a simple pendulum is simple harmonic in the limit of vanishingly small angle of oscillation. The motion of a mass on a spring is simple harmonic in the limit of vanishingly small amplitude of oscillation. In practice, such limiting conditions are rarely realised. Hence the exactly simple harmonic motions are rare.

Asymmetric (Non-simple Harmonic) Motion of Small Amplitude

In simple harmonic motion the amplitudes are equal on both sides of the equilibrium position. When this is not the case, the motion is not simple harmonic, however, small the amplitude may be. For example, a membrane loaded with a central mass on one side only vibrates with an -amplitude which is greater on one side of the undisplaced position than on the other. Its vibrations, therefore, are not simple harmonic.

Importance of S. H. M.

The study of simple harmonic motion is important for two reasons :

(i) There are a great many physical problems in mechanics, acoustics, optics, electricity and in atomic and molecular physics, in which the force on a system is approximately proportional to its displacement from some equilibrium position. In such cases, the resulting motion may be represented approximately by the simple harmonic model.

(ii) Even the complicated periodic motions occurring in physical problems can be represented as a combination of a number of simple harmonic motions having frequencies which are multiple of that of the complicated motion. The vibrating strings and membranes, the vibrations of atoms in solids, the electrical and acoustical oscillations in a cavity can be treated in this manner.

Energy of a Particle Performing S. H. M.

An oscillating particle possesses both types of energies; potential as wen as kinetic. It possesses potential energy on account of its displacement from the equilibrium position, and kinetic energy on account of its velocity. Naturally these energies vary during oscillation. Their sum is, however, conserved if no dissipative forces are present.

Potential Energy: Let us consider a particle of mass m executing S.H.M. Let x be its displacement from the equilibrium position at any instant t. Since in S. H. M. the force F acting upon the particle is proportional and opposite to the displacement x, we have

$$F = -kx,$$

where the constant k gives the force per unit displacement. We know that the force can also be expressed in terms of the potential energy U of the particle according as

$$F = -\frac{dU}{dx}.$$

Thus $$\frac{dU}{dx} = kx$$

Integrating, we get

$$U = \tfrac{1}{2}\ kx^2 + C \text{ (a constant).}$$

If we assume the potential energy zero in the equilibrium position, *i.e.* if U=0 at x=0, then C=0, Therefore

$$U = \tfrac{1}{2}\ kx^2.$$

But for S. H. M. we have

$$x = a \sin(\omega t + \phi), \qquad [\text{where } \omega^2 = k/m]$$

so that $$U = \tfrac{1}{2}\ ka^2\ (\omega t + \phi). \qquad ...(i)$$

This shows that the potential energy U varies with time t, attaining a maximum value $\frac{1}{2}\ ka^2$ when $\sin^2(\omega t + \phi)$=1, *i.e.* when x=±a.

Kinetic Energy: The kinetic energy K of the particle at an instant t is $\frac{1}{2}$ mu². Using the relation

$$x = a \sin(\omega t + \phi)$$

so that $$u = \frac{dx}{dt} = \omega a \cos(\omega t + \phi)$$

We obtain $$K = \frac{1}{2} mu^2 = \frac{1}{2} m\omega^2 a^2 \cos^2(\omega t + \phi)$$

New, $\omega^2 = k/m$.

$$\therefore \quad K = \frac{1}{2} ka^2 \cos^2(\omega t + \phi). \quad ...(ii)$$

This shows that the kinetic energy K also attains a maximum value $\frac{1}{2} ka^2$ when $\cos^2(\omega t + \phi) = 1$, *i.e.* when $x = 0$.

Total Energy: The total energy E of the particle is the sum of the kinetic and potential energies. Using eq. (i) and (ii), we obtain

$$E = K + U$$

$$= \frac{1}{2} ka^2 \cos^2(\omega t + \phi) + \frac{1}{2} ka^2 \sin^2(\omega t + \phi)$$

$$= \frac{1}{2} ka^2 [\cos^2(\omega t + \phi) + \sin^2(\omega t + \phi)]$$

$$= \frac{1}{2} ka^2$$

We thus see that the total energy is constant and has the value $\frac{1}{2} k\alpha^2$ This is also the maximum value of the potential or the kinetic energy. Thus at the time when the potential energy has its maximum value, the kinetic energy is zero, and vice-versa.

Further, $k = \omega^2 m$ and $\omega = \dfrac{2\pi}{T} = 2\pi n$, when n is the frequency of oscillations. Therefore

$$E = \frac{1}{2} ka^2 = \frac{1}{2} \omega^2 ma^2$$

$$= \frac{2\pi^2 ma^2}{T^2} = 2\pi^2 mn^2 a^2$$

Thus the total energy of the particle is directly proportional to the square of the amplitude (a^2) and inversely proportional to the square of the time period (T^2) or directly proportional to the square of the frequency (n^2). If the potential energy in the equilibrium position is not zero, then this value must be added to the above value of the total energy.

The distribution of energy versus displacement for S. H. M. is shown in figure below. The potential energy curve $U = -\frac{1}{2} kx^2 = \frac{1}{2} ka^2 \sin^2(\omega t + \phi)$ is a parabola with vertex at x=0, so that energy is stored in the oscillator both when x is positive and when it is negative. The kinetic energy curve $K = \frac{1}{2} mx^2 = \frac{1}{2}$ $ka^2 \cos^2(\omega t + \phi)$ is parabolic with respect to both x and a. The inversion of one curve with respect to the other means the $\pi/2$ phase difference between the displacement (related to U) and the velocity (related to k).

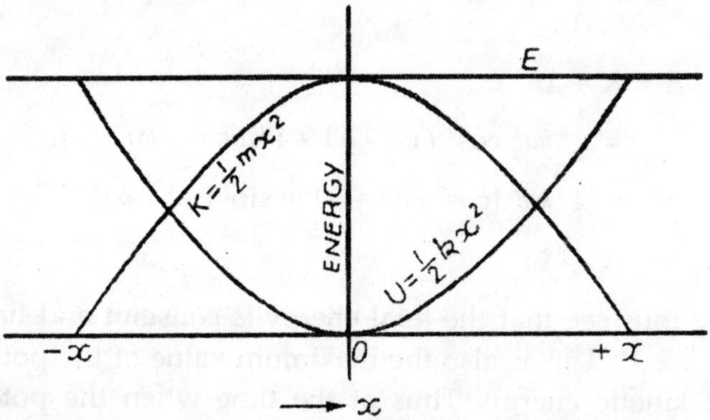

For any value of the displacement x the sum of the ordinates of both curves equals the total (constant) energy E, represented by the horizontal line.

Frequency of Energy-change in S. H. M.

The energy of a particle in S. H. M. (such as a simple pendulum) changes periodically from potential to kinetic and vice versa. At a displacement x, the energies are

$$U = \frac{1}{2} kx^2$$

and $$K = E - U = \frac{1}{2} k (a^2 - x^2).$$

The energy is entirety kinetic at the equilibrium position $(x=0)$ and entirely potential at the extreme position $(x=a)$.

Both U and K depend upon the square of the displacement x, that is, each acquires the same value twice in each oscillation.

Hence the frequency of energy-change is twice the frequency of oscillation.

Time-average of Kinetic and Potential Energies in S. H. M.

The instantaneous displacement of particle of mass m executing S. H. M. under a force-constant k is

$$x = a \sin(\omega t + \phi), \text{ where } \omega^2 = \frac{k}{m}.$$

The instantaneous kinetic energy is

$$K = \tfrac{1}{2} m \left(\frac{dU}{dx} \right)^2 = \tfrac{1}{2} m\omega^2 a^2 \cos^2(\omega t + \phi).$$

The time-average of the kinetic energy over a period T of the motion is

$$K_{av} = \frac{\int_0^T K\,dt}{T} = \frac{\tfrac{1}{2} m\omega^2 a^2 \int_0^{2\pi/\omega} \cos^2(\omega t + \phi)\,dt}{2\pi/\omega} \qquad \left[\because T = \frac{2\pi}{\omega}\right]$$

$$\frac{m\omega^3 a^2}{4\pi} \int_0^{2\pi/\omega} \frac{1 + \cos 2(\omega t + \phi)}{2}\,dt$$

$$= \frac{m\omega^3 a^2}{8\pi} \left[\int_0^{2\pi/\omega} dt \int_0^{2\pi/\omega} \cos 2(\omega t + \phi)\,dt \right]$$

$$= \frac{m\omega^3 a^2}{8\pi} \left[\frac{2\pi}{\omega} + 0 \right]$$

$$= \tfrac{1}{4} m\omega^2 a^2$$

$$= \tfrac{1}{4} ka^2. \qquad [\because \omega^2 = k/m]$$

The potential energy of the particle is

$$U = \tfrac{1}{2} kx^2 = \tfrac{1}{2} ka^2 \sin^2(\omega t + \phi).$$

The time-average of the potential energy over a period T of the motion is

$$U_{av} = \frac{\int_0^T U\,dt}{T} = \frac{\frac{1}{2}ka^2 \int_0^{2\pi/\omega} \sin^2(\omega t + \phi)}{2\pi/\omega}$$

$$= \frac{k\omega a^2}{4\pi} \int_0^{2\pi/\omega} \frac{1 - \cos 2(\omega t + \phi)}{2} dt$$

$$= \frac{k\omega a^2}{8\pi} [\int_0^{2\pi/\omega} dt - \int_0^{2\pi/\omega} \cos 2(\omega t + \phi)\, dt]$$

$$= \frac{k\omega a^2}{8\pi}\left[\frac{2\pi}{\omega} - 0\right]$$

$$= \tfrac{1}{4} ka^2.$$

Thus the time-average kinetic energy is equal to the time-average potential energy, and each is equal to $\frac{1}{4}$ *ka2*, which is half the total energy. This is a characteristic property of the harmonic oscillator and is not true for an harmonic oscillators.

Kinetic and Potential Energies

Let us now find out the position-averages of the kinetic and potential energies. The kinetic energy is

$$K = \tfrac{1}{2} m \left(\frac{dx}{dt}\right)^2 = \tfrac{1}{2} k \left(a^2 - x^2\right).$$

The position-average over the displacement from x=0 to $x=a$ is

$$K = \frac{\int_0^a K\,dx}{a} = \frac{\frac{1}{2}k \int_0^a \left(a^2 - x^2\right) dx}{a}$$

$$= \frac{1}{2}\frac{k}{a}\left[a^2x - \frac{x^3}{3}\right]_0^a = \frac{1}{2}\frac{k}{a}\left[a^2 - \frac{a^3}{3}\right]$$

$$= \frac{1}{3}ka^2$$

The potential energy is

$$U = \frac{1}{2}kx^2$$

The position-average is

$$\overline{U} = \frac{\int_0^a U\,dx}{a} = \frac{\frac{1}{2}k\int_0^a x^2dx}{a}$$

$$= \frac{1}{2}\frac{k}{a}\left[\frac{x^3}{3}\right]_0^a = \frac{1}{2}\frac{k}{a}\frac{a^3}{3}.$$

$$= \frac{1}{6}ka^2.$$

Thus the position-average kinetic energy ($\frac{1}{3}ka^2$) is *not* equal to the position-average potential energy ($\frac{1}{6}ka^2$).

Fractions of Energies

The equation for S. H. M. is

$$X = a \sin(\omega t + \phi) \qquad \text{(where } \omega^2 = k/m\text{)}$$

so that

$$u = \frac{dx}{dt} = \omega a \cos(\omega t + \phi) = \pm\sqrt{\left(\frac{k}{m}\right)}\sqrt{\left[a^2 - a^2\sin^2(\omega t + \phi)\right]}$$

$$= \pm\sqrt{(k/m)}\sqrt{(a^2 - x^2)}.$$

The total energy of the particle in S. H. M. is

$$E = K + U$$

$$= \tfrac{1}{2}\, mu^2 + \tfrac{1}{2}\, kx^2$$

$$= \tfrac{1}{2} k\, (a^2 - x^2) + \tfrac{1}{2} kx^2$$

$$= \tfrac{1}{2}\, ka^2 .$$

Now, when the displacement is one-half the amplitude, *i.e.* at $x=a/2$, we have

$$(U)_{a/2} = \tfrac{1}{2} kx^2 = \tfrac{1}{2} k \left(\frac{a}{2}\right)^2 = \tfrac{1}{8} ka^2 = \tfrac{1}{4} E .$$

and $$(K)_{a/2} = \tfrac{1}{2} k \left(a^2 - x^2\right) = \tfrac{1}{2} k \left(a^2 - \frac{a^2}{4}\right) = \frac{3}{8} ka^2 = \tfrac{3}{4} E .$$

Thus the potential energy is one-fourth and the kinetic energy is three-fourth of the total energy.

Similarly, we can show that when the displacement is one-third of the amplitude $(x—a/3)$, the kinetic energy is $\tfrac{1}{9} E$ and the potential energy is $\tfrac{8}{9}$ E.

Energy : Half Kinetic and Half Potential

Now, let x be the displacement when the energy is half potential and half kinetic. For this,

$$U = \tfrac{1}{2} E.$$

or $$\tfrac{1}{2} kx^2 = \tfrac{1}{2} \left(\tfrac{1}{2} ka^2\right)$$

$$\therefore \quad x = \frac{a}{\sqrt{2}}$$

Oscillations in a Potential Well

When the motion of a particle moving in a potential field

remains confined to a certain bounded region, then the particle is said to be in a 'potential well' or 'potential energy well'.

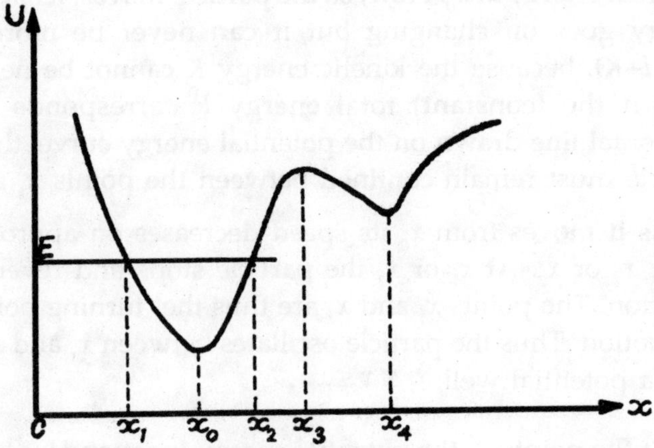

(Fig.)

Let us consider a particle moving under the influence of a one-dimensional force which depends only on the position x of the particle. The potential energy of the particle is also a function of the position only and can be expressed as $U(x)$. The negative derivative of the potential energy gives the force F acting on the particle, that is,

$$F = -\frac{dU}{dx}$$

The potential energy U may be any continuous function of the position x, such as shown in figure above. This is called the 'potential energy curve'. Its shape can give sufficient information regarding the motion of the particle.

At points x_0, x_3 and x_4 the slope of the curve, $\frac{dU}{dx}$, is zero so that the force is zero. A particle at rest at these points will remain at rest. However, the points x_0 and x_4 are the points of stable equilibrium, whereas the point x_3 is a point of unstable equilibrium.

Let us consider the particle at the point x_0. It can be displaced from x_0 only if its total energy E is more than the minimum potential energy at x_0. Now, as the particle moves, its potential energy goes on changing but it can never be more than E $(=U+K)$, because the kinetic energy K cannot be negative. Thus if the (constant) total energy E corresponds to the horizontal line drawn on the potential energy curve, then the particle must remain confined between the points x_1 and x_2.

As it moves from x_0 its speed decreases on approaching either x_1 or x_2. At x_1 or x_2 the particle stops and reverses its direction. The points x_1 and x_2 are thus the 'turning points' of the motion. Thus the particle oscillates between x_1 and x_2, and is in a potential well, $x_1 \leq x \leq x_2$.

At the point x_0 , the potential energy function $U(x)$ can be expanded into a Taylor's series :

$$U(x) = U_0 + \left(\frac{dU}{dx}\right)_0 (x - x_0) + \left(\frac{d^2U}{dx^2}\right)_0 \frac{(x - x_0)^2}{2} + \dots\dots\dots$$

where U_0 , $\left(\frac{dU}{dx}\right)_0$ are the values at the point $x=x_0$. If x_0 is a point of stable equilibrium, then $\left(\frac{dU}{dx}\right)_0$ is zero and $\left(\frac{d^2U}{dx^2}\right)_0$,

is a positive quantity, say C_2. Putting C_2, C_3,......... for $\left(\frac{d^2U}{dx^2}\right)_0$, $\left(\frac{d^3U}{dx^3}\right)_0$ ....we have

$$U(x) = U_0 + \frac{C_2}{2}(x - x_0)^2 + \frac{C_3}{3}(x - x_0)^3 + \dots\dots$$

If we shift the origin to x_0 , then we can put x in place of $x—x_0$. Then we have

$$U(x) = U_0 + \frac{C_2}{2} x^2 + \frac{C_3}{3} x^3 + \dots\dots$$

For small oscillations, the terms containing x^3 and higher can be ignored. Then

$$U(x) = U_0 + \tfrac{1}{2} C_2 x^2.$$

The U-x curve is now a parabola. Thus, for small-amplitude oscillations the potential well is parabolic.

Parabolic Potential Well: The force acting on the particle is given by

$$F = -\frac{dU(x)}{dx}$$

For a parabolic potential well, we have

$$U(x) = U_0 + \tfrac{1}{2} C_2 x^2$$

$$\therefore \quad F = -\frac{d}{dx}\left(U_0 + \tfrac{1}{2} C_2 x^2\right)$$

$$= -C_2 x$$

The force is thus proportional to the displacement from the equilibrium point and is directed towards it. This is the characteristic of S. H. M. Thus, for small oscillations, the motion in a potential well is simple harmonic.

The significance of a parabolic potential well lies in the fact that it is associated with simple harmonic motion. The concept can, therefore, be applied to a wide variety of systems.

Mass Attached to Horizontal Spring

Suppose a mass *m* attached to one end of a massless spring is free to move on a frictionless horizontal surface. Let x=0 be the equilibrium position of the mass at which the spring is unstretched. When the mass is displaced towards right through a distance x, then the spring exerts an elastic *restoring* force F on the mass. By Hocke's law,

$$F = -kx$$

where *k* is the force-constant of the spring. The negative sign indicates that F is directed towards left (opposite to displacement). (If we displace the mass towards left, then F would be directed towards right. When the displaced mass is released, it oscillates to and fro about the equilibrium position.

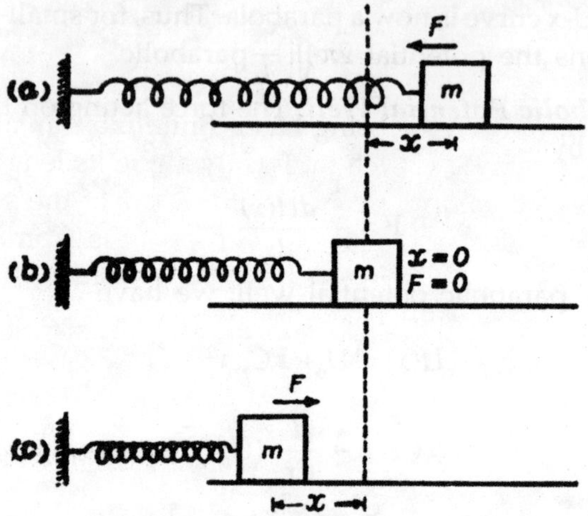

Fig.

At any instant during oscillation the force F is equal to the mass m multiplied by the instantaneous acceleration $\frac{dx^2}{dt^2}$.

That is

$$F = -kx = m\frac{d^2x}{dt^2}$$

or
$$\frac{d^2x}{dt^2} + \frac{k}{m}x = 0$$

or
$$\frac{d^2x}{dt^2} + \omega^2 x = 0$$

where $\omega^2 = k/m$. This is the differential equation of the simple harmonic motion of the mass. Its solution is

$$x = a\sin(\omega t + \phi),$$

where ω^2 and ϕ are constants. If t is replaced by $t + \frac{2\pi}{\omega}$, the displacement x will be the same. Hence the period of the motion is

$$T = \frac{2\pi}{\omega} = 2\pi\sqrt{\left(\frac{m}{k}\right)}.$$

Now, suppose the spring has a finite mass m_s. It $m_s << m$, then the spring will stretch uniformly along its length, that is, the displacements of the different elements of the spring will be proportional to their respective distances from the fixed end.

Let l be the length of the spring. Then its mass per unit length is m_s/l. Let ds be a small element of length at a distance s from the fixed end. Its mass is $(m_s/l)ds$.

When the displacement of the mass m is x, the displacement of the element ds will be $(s/l)x$ and the velocity will be $(s/l)\frac{dx}{dt}$. Hence the instantaneous kinetic energy of the element

$$= \frac{1}{2}\left(\frac{m_s}{l}\right)\left(\frac{s}{l}\frac{dx}{dt}\right)^2$$

$$= \frac{m_s}{2l^3}\left(\frac{dx}{dt}\right)^2 s^2\, ds.$$

The instantaneous kinetic energy of the whole spring is

$$= \frac{m_s}{2l^3}\left(\frac{dx}{dt}\right)^2 \int_0^l s^2\, ds = \frac{1}{6} m_s \left(\frac{dx}{dt}\right)^2.$$

The kinetic energy of the system (mass+spring) is

$$K = \frac{1}{2} m \left(\frac{dx}{dt}\right)^2 + m_s \left(\frac{dx}{dt}\right)^2$$

$$= \frac{1}{2}\left(\mathrm{m} + \frac{m_s}{3}\right)\left(\frac{dx}{dt}\right)^2.$$

This shows that the effective mass of the system is $\left(m + \frac{m_s}{3}\right)$. Hence the new time-period is

$$T = 2\pi\sqrt{\left(\frac{m + \frac{m_s}{3}}{k}\right)}.$$

Period of Oscillation of Mass on a Vertical Spring

Let us consider a massless spring of length l hanging vertically. When a mass m is attached to its lower end, its length extends say by x_0. The spring exerts an elastic - (restoring) upward force F on the mass m. By Hooke's law,

$$F = -kx_0$$

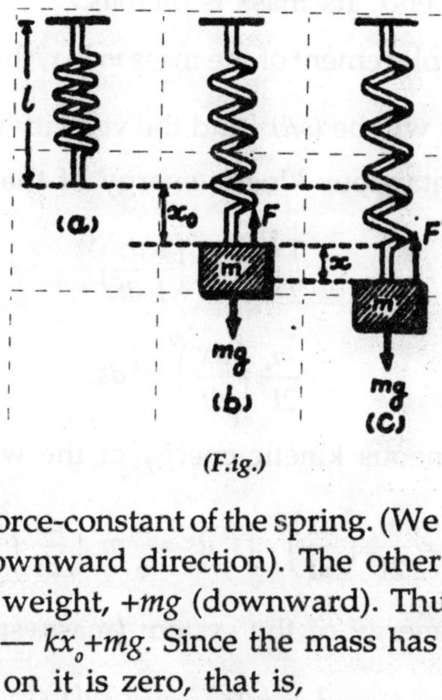

(F.ig.)

where k is the force-constant of the spring. (We take the positive force in the downward direction). The other force acting on the mass is its weight, $+mg$ (downward). Thus the total force acting on m is — kx_0+mg. Since the mass has no acceleration, the total force on it is zero, that is,

$$-kx_0+mg = 0$$

or $$x_0 = mg/k. \quad \text{...(i)}$$

This is the static equilibrium position of the mass m.

Now, suppose during oscillation, the total extension in the spring at any instant is x_o+x. The (upward) force exerted on *m* by the spring is then

$$F' = -k\,(x_o+x)$$

$$= -k\left(\frac{mg}{k}+x\right) \qquad \text{...[from eq. (i)]}$$

$$= -mg - kx.$$

The other force acting on the mass *m* is its weight, $+mg$ (downward). Therefore, the net force acting on the mass is

$$(-mg-kx)+mg = -kx.$$

This, by Newton's second law, must be *m* times the instantaneous acceleration *d2x/dt2* of the mass. That is

$$-kx = m\frac{d^2x}{dt^2}$$

or
$$\frac{d^2x}{dt^2} = -\frac{k}{m}x = -\omega^2 x,$$

where $\omega^2 = k/m$. This is the equation of the S. H. M. Thus the mass *m* executes S. H. M. about the position $x_o = mg/k$. The time-period is

$$T = \frac{2\pi}{\omega} = 2\pi\sqrt{\frac{m}{k}}. \qquad \text{...(ii)}$$

From eq. (i), we have $k = mg/x_0$. Therefore

$$T = 2\pi\sqrt{\frac{x_0}{g}}. \qquad \text{...(iii)}$$

Eq. (ii) and (iii) both can be used to determine the period of oscillation of the massless-spring system. Eq. (iii) has the advantage that we can find the period of oscillation simply by measuring the extension x_0 of the spring, without knowing the suspended mass and the force-constant of the spring.

Spring Cat in Two Halves: For a given mass hung from the spring, the extension x_0 is proportional to the unstretched length l of the spring. Therefore, for half length $(l/2)$ of the spring the extension will be $x_0/2$. Hence the new time-period will be

$$T = 2\pi\sqrt{\left(\frac{x_0/2}{g}\right)} = T/\sqrt{(2)},$$

or the new frequency is $\sqrt{(2)}$ times the old frequency.

Mass Suspended by Heavy Spring

Suppose a mass m is suspended from the end of a spring of mass m_s. If $m_s << m$, then the spring will stretch uniformly along its length. Let l be the length of the spring. Then its mass per unit length is m_s/l. Let us consider an element of length ds at a distance s from the fixed end of the spring. The mass of this element is $(m_s/l)\, ds$.

Let x be the instantaneous displacement of the mass m from its position of rest. Since the displacement is proportional to the distance from the fixed end, the displacement of the element considered will be $(s/l)\, x$, and its instantaneous velocity is $\left(\frac{s}{l}\right)\frac{dx}{dt}$. Hence the instantaneous kinetic energy of the element

$$= \frac{1}{2}\left(\frac{m_s}{l}\right) \left(\frac{s}{l}\frac{dx}{dt}\right)^2$$

$$= \frac{m_s}{2l^3}\left(\frac{dx}{dt}\right)^2 s^2\, ds.$$

If the spring is uniform, the total kinetic energy of the suspended mass and the spring is

$$K = \frac{1}{2}m\left(\frac{dx}{dt}\right)^2 + \frac{m_s}{2l^3}\left(\frac{dx}{dt}\right)^2 \int_0^l s^2\, ds$$

$$= \frac{1}{2} m \left(\frac{dx}{dt}\right)^2 + \frac{m_s}{2l^3}\left(\frac{dx}{dt}\right)^2 \frac{l^3}{3}$$

$$= \frac{1}{2}\left(m + \frac{m_s}{3}\right)\left(\frac{dx}{dt}\right)^2.$$

Now; when the mass is at a distance x from its position of rest, the elastic potential energy of the system is

$$U = \tfrac{1}{2}\, kx^2,$$

where k is the force-constant of the spring. Therefore, the total energy of the system is

$$E = K + U = \frac{1}{2}\left(m + \frac{m_s}{3}\right)\left(\frac{dx}{dt}\right)^2 + \frac{1}{2} kx^2 .$$

Since E remains constant, $\frac{dE}{dt} = 0$, that is

$$= \frac{1}{2}\left(m + \frac{m_s}{3}\right) 2\left(\frac{dx}{dt}\right)\frac{d^2x}{dt^2} + \frac{1}{2} k\,(2x)\frac{dx}{dt} = 0.$$

or

$$\left(m + \frac{m_s}{3}\right)\frac{d^2x}{dt^2} + kx = 0$$

or

$$\frac{dx}{dt} = -\frac{k}{m + (m_s/3)}\, x = -\,\omega^2 x,$$

where $\omega^2 = \frac{k}{m + (m_s/3)}$

Thus the motion is simple harmonic and its period is

$$T = \frac{2\pi}{\omega} = 2\pi\sqrt{\left(\frac{m + (m_s/3)}{k}\right)}$$

The frequency is

$$n = \frac{1}{T} = \frac{1}{2\pi}\sqrt{\left(\frac{k}{m + (m_s/3)}\right)}.$$

If the spring be massless, the period is $T = 2\pi\sqrt{m/k}$. Thus when the mass of the spring is taken into account, the period is increased or the frequency is decreased.

Calculation of Period of S.H.M. of Centre of Mass of the Cylinder

Let m be the mass, r the radius, and I the moument of inertia of the cylinder about its own axis. Let k be the force-constant of the spring. When the cylinder is pulled and released, it begins to roll without slipping, and the centre of mass of the cylinder executes S. H. M.

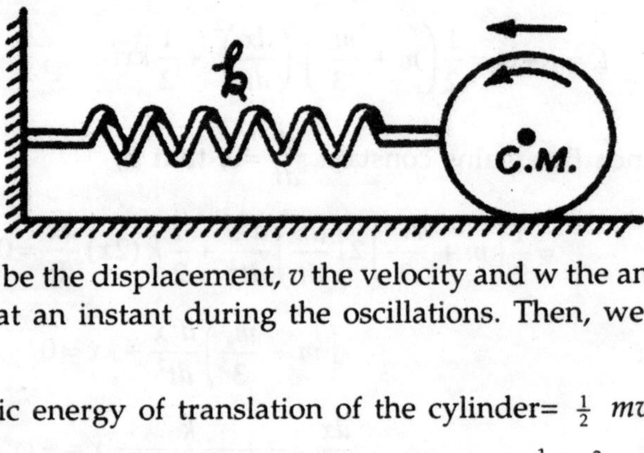

Let x be the displacement, v the velocity and w the angular velocity at an instant during the oscillations. Then, we have have

kinetic energy of translation of the cylinder= $\frac{1}{2} mv^2$,

kinetic energy of rotation of the cylinder= $\frac{1}{2} I\omega^2$,

and potential energy of the spring = $\frac{1}{2} kx^2$.

By the conservation of energy, we have

$$\frac{1}{2} mv^2 + \frac{1}{2} I\omega^2 + \frac{1}{2} kx^2 = \text{constant.}$$

But $I = \frac{1}{2} mr^2$ and $\omega = \frac{v^2}{r^2}$ Thus

$$\frac{3}{2} mv^2 + \frac{1}{2}\left(\frac{1}{2} mr^2\right)\frac{v^2}{r^2} + \frac{1}{2} kx^2 = \text{constant}$$

or $$\frac{3}{4} mv^2 + \frac{1}{2} kx^2 = \text{constant.}$$

Differentiating it, we get

$$\tfrac{3}{2} mv \frac{dv}{dt} + kx \frac{dx}{dt} = 0$$

or
$$\tfrac{3}{2} m \left(\frac{dv}{dt}\right) \frac{d^2x}{dt^2} + kx \frac{dx}{dt} = 0$$

or
$$\tfrac{3}{2} m \frac{d^2x}{dt^2} + kx = 0$$

or
$$\frac{d^2x}{dt^2} = -\frac{2k}{3m} x = -\omega^2 x$$

where $\omega^2 = 2k/3m$.

This is the equation of S. H. M. of period

$$T = \frac{2\pi}{\omega} = 2\pi\sqrt{\left(\frac{2k}{3m}\right)}$$

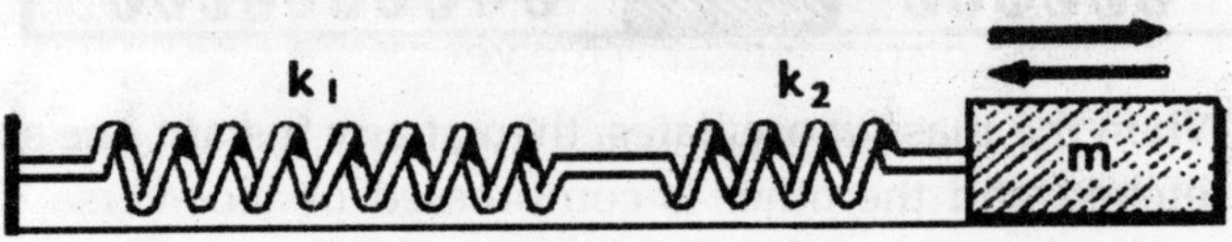

When the mass m oscillates, both the springs are stretched (or compressed) simultaneously. Let x be the displacement of m from its equilibrium position at an instant, and x_1, x_2 the extensions (or contractions) in the length of the springs k_1, k_2 respectively at this instant. Then

$$x = x_1 + x_2 . \qquad \text{...(i)}$$

The same elastic (restoring) force F is developed in each spring. By Hooke's law, we have

$$F = -k_1 x_1 = -k_2 x_2 .$$

Therefore, from eq. (i), we have

$$x = -\frac{F}{k_1} - \frac{F}{k_2}$$

or
$$F = -\frac{k_1 k_2}{k_1 + k_2} x.$$

Thus the force-constant of the system is $\frac{k_1 k_2}{k_1 + k_2}$. Hence the period of motion is

$$T = 2\pi \sqrt{\left(\frac{k}{m}\right)} = 2\pi \sqrt{\left[\frac{(k_1 k_2)\, m}{k_1 + k_2}\right]}.$$

The frequency is

$$n = \frac{1}{T} = \frac{1}{2\pi} \sqrt{\left[\frac{k_1 k_2}{(k_1 + k_2)\, m}\right]}.$$

Calculation of Period and Frequency

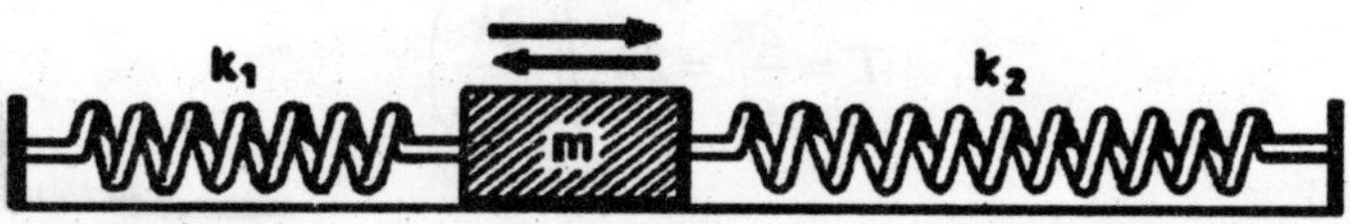

When the mass m oscillates, then at any instant one spring is stretched and the other is compressed or vice-versa. Let x be the displacement of m from its equilibrium position at an instant. This will also be the extension in one and the compression in the other spring.

The restoring force developed in the two springs are
$$F_x = - k/x \text{ and } F_2 = - k_3 x$$
and act on the mass along the same direction. Therefore, the total restoring force is
$$F = F_1 + F_2 = -(k_1 + k_2)\, x.$$

Thus the force-constant of the system is $(k_1 + k_2)$, and hence its period is

$$T = 2\pi \sqrt{\left(\frac{m}{k}\right)} = 2\pi \sqrt{\left(\frac{m}{k_1 + k_2}\right)},$$

and the frequency is

$$n = \frac{1}{2\pi}\sqrt{\left(\frac{k}{m}\right)} = \frac{1}{2\pi}\sqrt{\left(\frac{k_1 + k_2}{m}\right)}.$$

Simple Pendulum

A simple pendulum is an ideal body consisting of a point mass suspended by a weightless, inextensible cord from a rigid support.

In practice, a small metallic bob suspended by a cotton thread is used as a simple pendulum.

Motion of Simple Pendulum: Let *m* be the mass of the bob and / the length of the thread. In the equilibrium position, the bob hangs vertically below the point of suspension. When the bob is displaced slightly from this equilibrium position and released, it oscillates in a vertical plane under gravity. Let θ (in radian) be the angular displacement of the pendulum from its equilibrium position at an instant during the oscillation. The corresponding displacement along the arc is $x = l\theta$.

The forces acting on m are *mg*, its weight acting vertically downward, and T the tension in the thread.

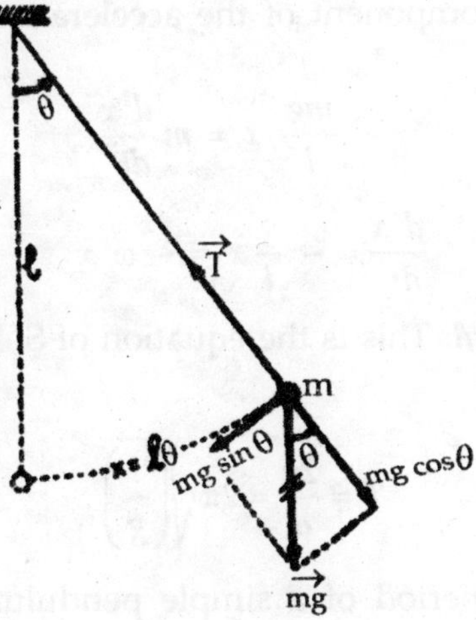

The force mg can be resolved into two components; a radial component $mg \cos \theta$ along the thread and a tangential component $mg \sin \theta$ at right angles to it. The tangential component $mg \sin \theta$ is the *restoring* force F acting on m tending to return it to the equilibrium position. Thus

$$F = -mg \sin \theta .$$

The minus sign indicates that the force is 'restoring' (opposite to the displacement θ. We see that the (tangential) restoring force is *not* proportional to the angular displacement θ, but it is proportional to sin θ. Hence the motion of the pendulum is *not* simple harmonic. But if the angle θ is sufficiently small, then $\sin \theta = \theta$ (in radian) so that

$$F = -mg\theta = -mg\frac{x}{l} - = -\frac{mg}{l}x.$$

Now the restoring force is proportional to the displacement and the motion is simple harmonic.

Now, by Newton's second law, the (tangential) restoring force on m must be equal to the product of the mass m and the tangential component of the acceleration, $\frac{d^2x}{dt^2}$ Thus

$$-\frac{mg}{l}x = m\frac{d^2x}{dt^2}$$

or $$\frac{d^2x}{dt^2} = -\frac{g}{l}x = -\omega^2 x$$

where $\omega^2 = g/l$. This is the equation of S. H. M. with time-period

$$T = \frac{2\pi}{\omega} = 2\pi\sqrt{\left(\frac{l}{g}\right)}.$$

This is the period of a simple pendulum, provided the

amplitude of oscillations is sufficiently small. It is independent of the mass of the bob, as well as of the amplitude.

Large Amplitude: When the amplitude of motion of the simple pendulum is large, the approximation $\sin\,\theta = \theta$ does not hold and the pendulum deviates from the simple harmonic behaviour. Then the time-period is given by

$$T = 2\pi\sqrt{\left(\frac{l}{g}\right)}\left(1+\frac{1}{2^2}\sin^2\frac{\theta_m}{2}+\frac{1}{2^2}\frac{3^2}{4^2}\sin^4\frac{\theta_m}{2}+\ldots\right),$$

where θ_m is the angular amplitude. Thus with increasing amplitude the period increases.

An Expression for the Tension in the String of a Simple Pendulum

Let m be the point-mass and l the length of the pendulum. We have seen that the radial forces acting on the mass m are the tension T in the string and the component $mg\cos\,\theta$ of the weight. These forces provide the centripetal force acting on m which is moving in a circle of radius l. Thus

$$T = mg\cos\,\theta = \frac{mv^2}{l},$$

where v is velocity of m when the angular displacement is θ.

$$\therefore T = mg\cos\,\theta = \frac{mv^2}{l}. \qquad \text{(i)}$$

Now, the equation of motion of the pendulum is

$$\frac{d^2x}{dt^2} = -\frac{g}{l}x,$$

where x is the displacement along the circular path. Multiplying it by 2 (dx/dt), we have

$$2\left(\frac{dx}{dt}\right)\frac{d^2x}{dt^2} = -\frac{l}{g}2x\frac{dx}{dt}.$$

Integrating, we get

$$\left(\frac{dx}{dt}\right)^2 = -\frac{l}{g}x^2 + C \text{ (constant)}$$

or

$$v^2 = -\frac{l}{g}x^2 + C$$

where $v(=dx/dt)$ is the velocity of m. We know that the velocity of the mass is zero at the extreme positions. That is, at $x=a$ (amplitude), $v=0$. Therefore $C=(g/l)\,a^2$.

$$\therefore v^2 = \frac{l}{g}\left(a^2 - x^2\right).$$

Let θ_m be the angular amplitude. Then $a—l\,\theta_m$ and $x=l\,\theta$. Thus

$$v^2 = \frac{l}{g}\left(l^2\theta_m{}^2 - l^2\theta^2\right) = gl\left(\theta_m{}^2 - \theta^2\right).$$

Substituting this value of v^2 in eq. (i), we get

$$T = mg\cos\theta + mg\left(\theta_m{}^2 - \theta^2\right).$$

If θ is small, then $\cos\theta = 1 - \frac{\theta^2}{2}$. Therefore

$$T = mg\left(1 - \frac{\theta^2}{2}\right) + mg\left(\theta_m{}^2 - \theta^2\right)$$

or

$$\boxed{T = mg\left(1 - \frac{3\theta_m{}^2}{2} + \theta_m{}^2\right).}$$

This is the expression for the tension in the string.

The tension is a minimum when $\theta = \theta_m$, *i.e.* when the pendulum is in an extreme position. Thus

$$T_{min} = mg\left(1 - \frac{3\theta_m{}^2}{2} + \theta_m{}^2\right) = mg\left(1 - \frac{\theta_m{}^2}{2}\right).$$

The tension is a maximum when $\theta = 0$ *i.e.* when the pendulum is passing through its equilibrium position. Thus

$$T_{max} = mg\,(1 + \theta_m^{\,2}).$$

Angular Simple Harmonic Motion of a Torsion Pendulum

Let a body, say a disc, be hung by a long and thin vertical wire whose upper end is rigidly clamped. Let the disc be given a slight rotation in the horizontal plane. The wire is twisted and, due to its elasticity, exerts a restoring torque on the distending to return it to the initial position. When the disc is released, the restoring torque produces an angular acceleration in the disc in a direction opposite to that of the twist. The disc therefore returns to its equilibrium position, crosses it (due to inertia), and rotates further in the opposite direction. The wire is now twisted in the opposite direction. A restoring torque is again exerted on the disc which arrests its motion and makes it return. The whole phenomenon is then repeated.

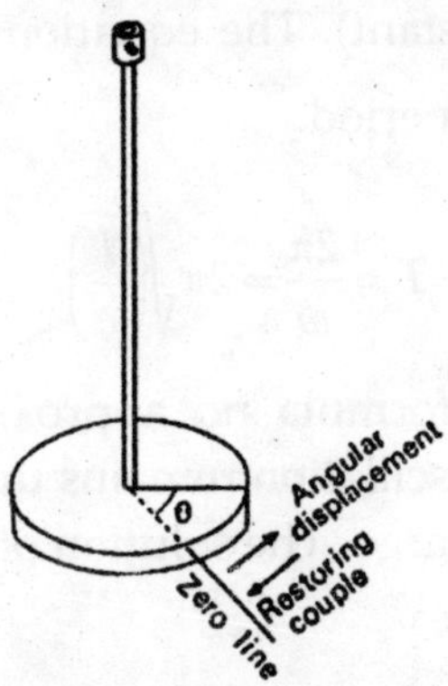

Thus the disc oscillates in the horizontal plane about the wire as axis. These oscillations are called 'torsional oscillations' and the system is called a 'torsion pendulum'.

Expression for the Periodic Time: Suppose, when the disc is oscillating, the twist at the lower end of the wire at an instant is θ. The restoring torque I at this instant will be given by

$$I = -c\theta,$$

where c is the restoring torque per unit twist. The minus sign is put as the torque is directed opposite to the twist θ. By Newton's second law (for angular motion), the torque acting on the disc must be equal to the product of the moment of inertia I of the disc about the axis of oscillation and the instantaneous angular acceleration $\frac{d^2\theta}{dt^2}$. That is

$$-c\theta = I\frac{d^2\theta}{dt^2}$$

or

$$\frac{d^2\theta}{dt^2} + \frac{c}{I}\theta = 0$$

or

$$\frac{d^2\theta}{dt^2} + \omega^2\theta = 0,$$

where $\omega^2 = \frac{c}{I}$ (a constant). The equation represents a simple harmonic motion of period.

$$T = \frac{2\pi}{\omega} = 2\pi\sqrt{\left(\frac{I}{c}\right)}.$$

In deriving this formula no approximation is required. Hence the period of oscillation remains unchanged even if the angular amplitude is large. The solution of the equation maybe written as

$$\theta = \theta_0 \sin(\omega t + \phi),$$

where θ_0 is angular amplitude and ϕ is initial phase.

Compound Pendulum

Any rigid body which is so mounted that it can swing in a vertical plane about a horizontal axis passing through it is called a 'physical' or 'compound' pendulum.

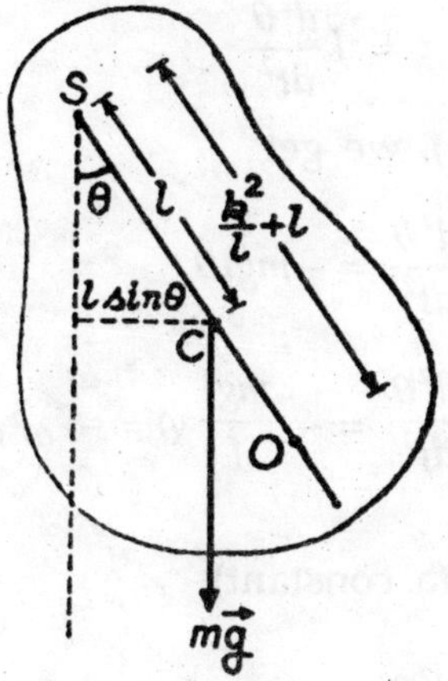

The above figure represents the vertical section of an irregular rigid body pivoted at a point S in it. In the equilibrium position, the centre of mass C of the body lies vertically below *S*. When the body is displaced to one side and then released, it oscillates m the vertical plane containing *S* and *C* about a horizontal axis perpendicular to this plane and passing through *S*. It is now acting as a compound pendulum.

Motion of Compound Pendulum: Let *m* be the mass of the body and *I* the distance of C from S. Suppose at some instant during the oscillation, the body makes an angle θ with the vertical. At this instant the moment of its weight *mg* (acting vertically at C*)* about the axis of oscillation through *S* is *mg* (l sin θ*)*. This is the restoring torque τ (say) at this instant tending to bring the body to its equilibrium position Thus

$$\tau = — mgl \sin\theta .$$

The negative sign is used since the torque acts in the direction of θ decreasing. If the angle θ is small, then sin $\theta = \theta$ (in radian) and so

$$\tau = —mgl\,\theta . \qquad \text{...(i)}$$

If *I* be the moment of inertia of the body about the axis through *S*, and $d^2\theta/dt^2$ the instantaneous angular acceleration, then the torque acting on the body at this instant must be given by

$$\tau = I\frac{d^2\theta}{dt^2}$$

From (i) and (ii), we get

$$I\frac{d^2\theta}{dt^2} = -mgl\theta$$

or
$$\frac{d^2\theta}{dt^2} = -\frac{mgl}{I}\theta = -\omega^2\theta,$$

where $\omega^2 = \frac{mgl}{I}$ (a constant).

$$\therefore \quad \frac{d^2\theta}{dt^2} \propto \theta.$$

that is, the angular acceleration is proportional to the angular displacement. The motion is, therefore, *simple harmonic* and its time-period is given by

$$T = \frac{2\pi}{\omega} = 2\pi\sqrt{\left(\frac{I}{mgl}\right)}.$$

Let k be the radius of gyration of the pendulum about an axis passing through its centre of mass C and perpendicular to the plane of oscillation. Then its moment of inertia about this axis will be mk^2. Therefore, the moment of inertia I about the axis through S will be

$$I = mk^2 + ml^2 \text{ (theorem of parallel axes)}$$

$$\therefore \quad T = 2\pi\sqrt{\left(\frac{mk^2+ml^2}{mgl}\right)}.$$

or
$$T = 2\pi\sqrt{\left(\frac{\frac{k^2}{l}+l}{g}\right)} \qquad \text{...(iii)}$$

This is the expression for the time-period. Comparing it with the period of a simple pendulum

$$T = 2\pi\sqrt{\left(\frac{L}{g}\right)},$$

we see that the period of a compound pendulum is the same as that of a simple pendulum of length $L = \left(\frac{k^2}{l} + l\right)$. Thus $\left(\frac{k^2}{l} + l\right)$ is the "length of the equivalent simple pendulum."

Let us suppose that the whole mass of the compound pendulum is concentrated at a point *O* on *SC* produced such that $So = \left(\frac{k^2}{l} + l\right)$. Thus it would be like a simple pendulum of the same period. The point O which is at a distance equal to the length of the equivalent simple pendulum from S is called the "centre of oscillation" corresponding to the centre of suspension S.

Interchangeability of the Centres of Suspension and Oscillation: When the body is suspended from the centre of suspension S, its period of oscillation is given by

$$= 2\pi\sqrt{\left(\frac{\frac{k^2}{l} + l}{g}\right)},$$

where *l* is the distance of *S* from *C*.

Let the body be now inverted and suspended from the centre of oscillation *O*. The distance of *0* from C is k^2/l. The new period of oscillation *T′* is therefore obtained by substituting k^2/l for *l* in the above expression *i.e.*

$$T' = 2\pi\sqrt{\left(\frac{\frac{k^2}{k^2/l}+k^2/l}{g}\right)} = 2\pi\sqrt{\left(\frac{l+\frac{k^2}{l}}{g}\right)}.$$

Hence T = T′.

Thus the periods of oscillation about S and O are equal, i.e. the centres of suspension and oscillation are interchangeable.

Maximum and Minimum Periods of Compound Pendulum: The period T is given by

$$T^2 = \frac{4\pi^2}{g}\left[\frac{k^2}{l}+l\right].$$

Differentiating with respect to l, we get

$$2T\frac{dT}{dl} = \frac{4\pi^2}{g}\left[-\frac{k^2}{l^2}+1\right].$$

T will be a maximum or a minimum when $\frac{dT}{dl}=0$, that is, when

$$-\frac{k^2}{l^2}+1=0$$

or

$$\frac{k^2}{l^2}=1$$

or

$$l=\pm k.$$

With this value of l, the value of $\frac{d^2T}{dl^2}$ is positive. Hence the period T will be a 'minimum' when $l=\pm k$.

$$T_{min} = 2\pi\sqrt{\left(\frac{2k}{g}\right)}.$$

If the pendulum be suspended at its centre of mass (l=0), then

$$T = \infty .$$

This is the maximum period. It means that the pendulum suspended at its centre of mass will not oscillate.

Determination of g: To determine g by means of a compound pendulum, we have to find two points in the pendulum situated on a line on opposite sides of its centre of mass at *unequal* distances from it such that the periods of oscillation about them are equal. If one of the them is the centre of suspension then the other is the corresponding centre of oscillation. If L be the distance between them, which is the length of equivalent simple pendulum, then we have

$$T = 2\pi\sqrt{\left(\frac{L}{g}\right)}.$$

Thus, g can be worked out.

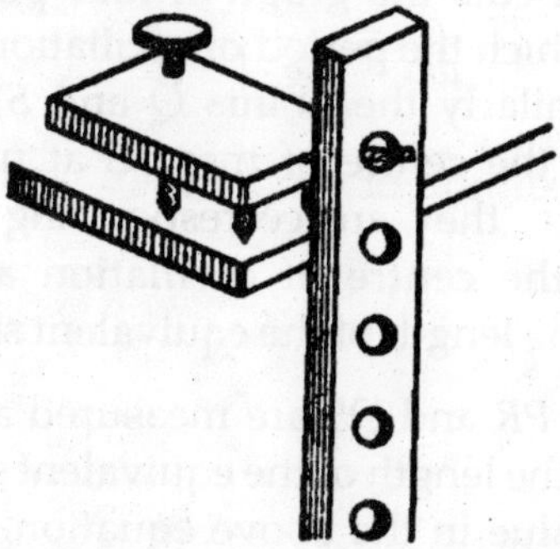

The pendulum is taken in the form of a bar with equidistant holes drilled along its length such that the centres of the holes are collinear with the centre of mass of the bar. The bar is suspended vertical on a knife-edge passing through a hole near its end. It is now slightly displaced to one side and released, so that it begins to oscillate in a vertical plane. The period of oscillation is obtained by noting time for some fifty oscillations. In a similar way, the period of oscillation is determined by placing the knife-edge in each hole turn by turn.

The period of oscillation T *is* plotted against the distance of knife-edge from the end of the bar. The graph so obtained has two parts. These parts are symmetrical about a line which intersects the x-axis at C, where C corresponds to the centre of mass of the bar about which the period is infinite.

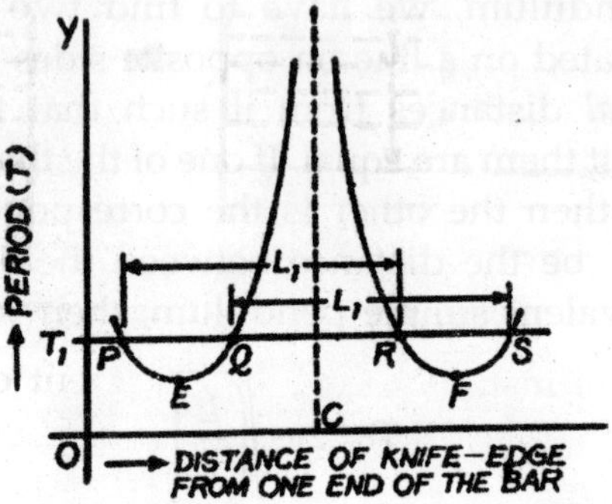

Now a straight line is drawn the *x*- axis and intersecting the *y*-axis at T_1. It cuts the graph at four points *P, Q, R* and *S*, about each of which the period of oscillation is T_1. The points *P* and *R* (and similarly the points *Q* and *S*) are situated on opposite sides of the centre of mass C at unequal distances from it. There-fore, they are corresponding to the centre of suspension and the centre of oscillation and the distance between them is the length of the equivalent simple pendulum.

The distances *PR* and *QS* are measured and their mean is found out. This is the length of the equivalent simple pendulum L_1. Putting this value in the above equation, the value of *g* is calculated. Similarly, we can take other lines parallel to x-axis corresponding to other periods of oscillation. The mean of the values of *g* so obtained is calculated.

Oscillations of a Floating Cylinder

Let us conside: a cylinder of mass *m* and cross-sectional area *A* floating vertically in a liquid of density *P*. Let *l* be the *immersed* length of the cylinder. At equilibrium, the weight of the cylinder is equal to the upthrust of the displaced liquid ($mg=Al\rho g$). When the cylinder is slightly pressed down and released, it oscillates up and down.

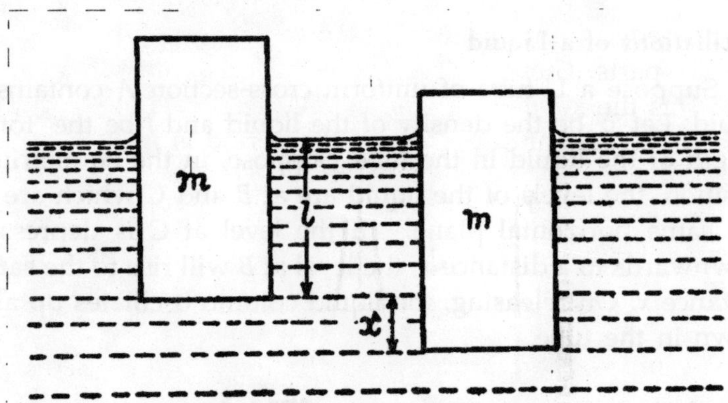

Let x be the instantaneous displacement of the cylinder. Then the mass of the further displaced liquid will be $Ax\rho$ and the corresponding upthrust will be $Ax\rho g$. This is the *restoring* force F on the cylinder, that is,

$$F = -Ax\rho g.$$

The force is directed opposite to the displacement x.

By Newton's second law, the force F equals the product of mass m and acceleration d^2x/dt^2 of the cylinder. Thus

$$-Ax\rho g = m\frac{d^2x}{dt^2}$$

or
$$\frac{d^2x}{dt^2} = -\frac{A\rho g}{m}x = -\omega^2 x,$$

where $\omega^2 = A\rho g/m$. This represents a S.H. M. of period

$$T = \frac{2\pi}{\omega} = 2\pi\sqrt{\left(\frac{m}{A\rho g}\right)}.$$

But $m = Al\rho$.

$$\therefore \quad T = 2\pi\sqrt{\left(\frac{l}{g}\right)}$$

Oscillations of a Liquid

Suppose a U-tube of uniform cross-section A contains a liquid. Let ρ be the density of the liquid and l be the 'total' length of the liquid in the tube. Suppose, in the equilibrium position, the levels of the liquid are at B and C which are in the same horizontal plane. ' If the level at C is depressed downwards to a distance x, the level at B will rise to the same distance x. On releasing, the liquid column oscillates up and down in the tube.

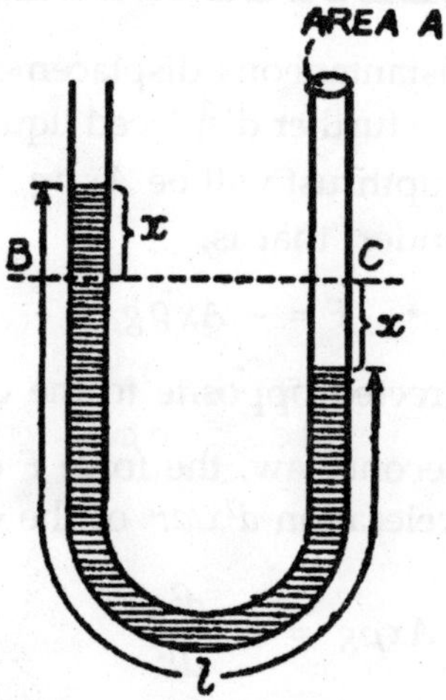

Let x be the instantaneous displacement of the liquid level below C. Then the length of the extra column in the left-hand limb will be $2x$. The weight of this extra column is $2xA\rho g$ and this is the restoring force F acting on the liquid. That is,

$$F = -2\,xA\rho g.$$

By Newton's law, this equals the product of the mass m of the liquid and its acceleration d^2x/dt^2. Thus

$$-2xA\rho g = m\frac{d^2x}{dt^2}$$

or $$\frac{d^2x}{dt^2} = -\frac{2A\rho g}{m}x = -\omega^2 x,$$

where $\omega^2 = \dfrac{2A\rho g}{m}$. This represents a S. H. M. of period

$$T = \frac{2\pi}{\omega} = 2\pi\sqrt{\left(\frac{m}{2\,A\rho g}\right)}$$

But $m = lA\rho$.

$$\therefore \quad T = 2\pi\sqrt{\left(\frac{l}{2g}\right)}.$$

Transverse Motion of a Mass between two Springs

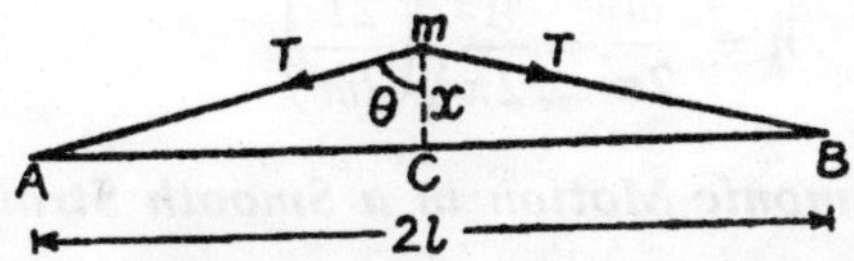

Suppose a mass m is suspended between two points A and B by means of two identical springs. Let T be the tension in each spring and $2l$ the horizontal distance between the points A and B. When the mass m is slightly displaced transversely and then released, it begins to oscillate up and down.

Let x be the instantaneous displacement of the mass m during oscillation, Suppose x is so small that the length l of each spring and the tension T in it remain unchanged.

The restoring force on the particle is the sum of the transverse components of the tensions T, T. Since the two springs are identical, the two transverse components are equal. Therefore, the restoring force acting upon the particle is given by

$$F = -2T\cos\theta$$

$$= -2T\left(\frac{x}{l}\right)$$

$$= -\frac{2T}{l}x,$$

the negative sign is used since the force acts in the direction of x decreasing.

By Newton's second law, the force F is the product of the mass m and its acceleration $d2x/dt^2$, that is,

$$-\frac{2T}{l}x = m\frac{d^2x}{dt^2}$$

or
$$\frac{d^2x}{dt^2} = -\frac{2T}{lm}x = -\omega^2 x,$$

where $\omega^2 = 2T/lm$. This represents a S. H. M. of frequency

$$n = \frac{\omega}{2\pi} = \frac{1}{2\pi}\sqrt{\left(\frac{2T}{lm}\right)}$$

Simple Harmonic Motion in a Smooth Straight Tunnel

Let O be the centre of the earth and P the position of the particle at any instant. Let $OP = r$. The (gravitational) force on the particle is that due to a sphere of radius r, since the outer shell exerts no force on the particle.

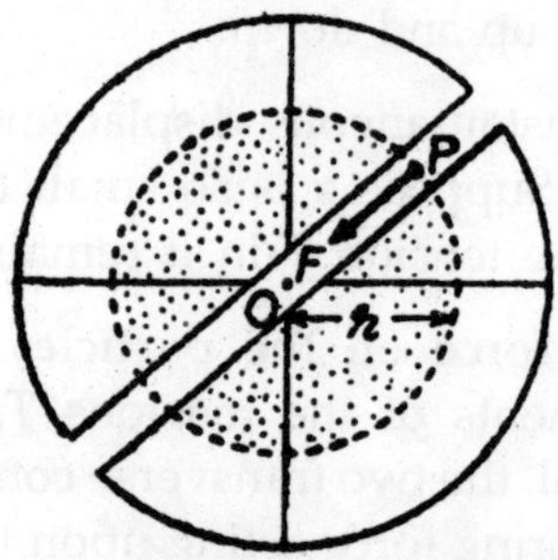

If ρ be the density of the earth, the mass of the inner solid sphere would be $\frac{4}{3}\pi r^3\rho$. Therefore the force on the particle P of mass m is

$$F = G\frac{\left(\frac{4}{3}\pi r^3\rho\right)m}{r^2}.$$

The minus sign indicates that the force is attractive and

directed towards the centre of the earth. The acceleration of P directed towards O is

$$a = \frac{F}{m} = -\tfrac{4}{3}\pi G\rho r = -\omega^2 r$$

where $\omega^2 = \tfrac{4}{3}\pi G\rho$ (constant). The acceleration is thus proportional to the displacement r but oppositely directed. Hence the particle oscillates about O in simple harmonic motion. The period of this S. H. M. is

$$T = \frac{2\pi}{\omega} = 2\pi\sqrt{\left(\frac{3}{4\pi G\rho}\right)} = \sqrt{\left(\frac{3\pi}{G\rho}\right)}.$$

We know that.

$$\rho = \frac{3g}{4\pi RG},$$

where R is the radius of the earth and g is acceleration due to gravity. Hence the period of motion may also be written as

$$T = 2\pi\sqrt{\left(\frac{R}{g}\right)}.$$

Helmholtz Resonator

It consists of a large spherical glass vessel having a small cylindrical neck A and a narrow stem B just opposite to A. When the neck A is directed towards a sound source, the air in the vessel vibrates with a natural frequency depending upon the volume of the vessel and the dimension of its neck.

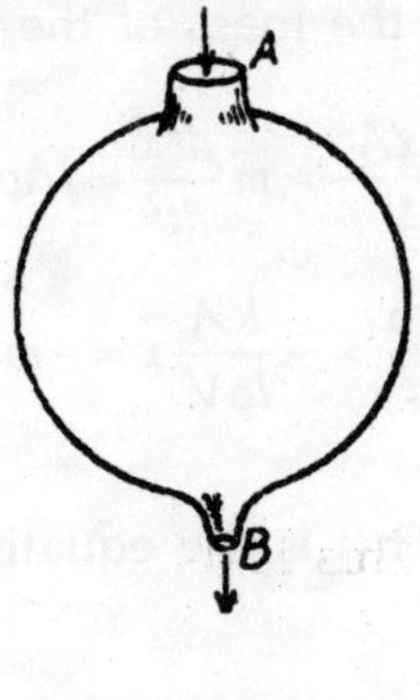

Let l be the length of the neck, A its area of cross-section and V the volume of the vessel. The mass of the air within the neck is

$$m = lA\rho$$

where ρ is the density of air. When sound waves enter the neck, the air in it acts like a piston, alternately compressing and ratifying the air inside the vessel. If the aïr within the neck moves inward through a small distance x then the air inside the vessel is compressed by an amount v given by

$$v = xA.$$

The resulting increase in pressure, p, is given by Hooke's law as

$$K = -\frac{\rho}{v/V}\left(\text{bulk modulus} = \frac{\text{volume stress}}{\text{volume strain}}\right)$$

or

$$p = -K\frac{v}{V}.$$

Therefore the (restoring) force on the air in the neck

$$F = \rho A = -\frac{KvV}{V} = -\frac{KxA^2}{V}.$$

But by Newton's second law

$$F = m\frac{d^2x}{dt^2}.$$

where $m\,(=lA\rho)$ is the mass of the air in the neck. Hence

$$-\frac{KxA^2}{V} = m\frac{d^2x}{dt^2} = lA\rho\frac{d^2x}{dt^2}$$

$$\frac{d^2x}{dt^2} = -\frac{KA}{l\rho V}x = -\omega^2 x,$$

where $\omega^2 = \dfrac{KA}{l\rho V}$. This is the equation of S. H. M. whose frequency is given by

$$n = \frac{\omega}{2\pi} = \frac{1}{2\pi}\sqrt{\left(\frac{KA}{l\rho V}\right)}$$

Now $\sqrt{\left(\frac{K}{\rho}\right)} = v$, the velocity of sound in air.

$$\therefore n = \frac{v}{2\pi}\sqrt{\left(\frac{A}{lV}\right)}$$

This is the natural frequency of the air of the resonator. If a sound wave is incident on the neck *A*, then the air will resonate only if its natural frequency is equal to the frequency of the incident wave. To detect the resonance easily, the stem *B* is pressed into the ear.

Equation for Simple Harmonic Oscillators

(i) Simple Pendulum:

$$\frac{d^2x}{dt^2} + \omega^2 x = 0;\ \omega^2 = \frac{g}{l} = \frac{\text{acceleration of gravity}}{\text{langht}}$$

(ii) Compound Pendulum:

$$\frac{d^2\theta}{dt^2} + \omega^2\theta = 0;\ \omega^2 = \frac{mgl}{I},$$

where *m* is the mass, *I* is the moment of inertia of the pendulum about the axis of suspension and *l* is the distance of the centre of mass from the centre of suspension.

(iii) Torsional Pendulum:

$$\frac{d^2\theta}{dt^2} + \omega^2\theta = 0;\ \omega^2 = \frac{c}{I} = \frac{\text{torsional constant}}{\text{moment of inertia}}$$

(iv) L-C Circuit:

$$\frac{d^2q}{dt^2}+\omega^2 q=0;\quad \omega^2=\frac{I}{LC}.$$

(v) Mass-Spring System:

$$\frac{d^2x}{dt^2}+\omega^2 x=0;\qquad \omega^2=\frac{k}{m}=\frac{\text{force constant}}{\text{moment of inertia}}.$$

(vi) Hehmholtz Resonator :

$$\frac{d^2x}{dt^2}+\omega^2 x=0;\qquad \omega^2=\frac{KA}{l\rho V},$$

where l and A are the length and cross-section f the neck of the resonator, V is the volume of the resonator, K and P are the bulk modulus and density of air.

Problems

1. t = 0, x = 0, A = vo particle is executing simple harmonic motion along the x-axis. Obtain expressions for its position, velocity as functions of time, assuming the initial conditions

Solution. Particle in S.H.M.: Let x be the displacement of the particle from its equilibrium position at any instant t. By definition of S.H.M., the instantaneous force acting upon the particle is given by

$$F = -kx,$$

where k is the proportionality factor. But by Newton's second law

$$F=m\frac{d^2x}{dt^2},$$

where m is the mass of the particle and d^2x/dt^2 is the instantaneous acceleration. Thus

$$m\frac{d^2x}{dt^2}=-kx$$

or
$$\frac{d^2x}{dt^2} = -\omega^2 x,$$

where $\omega^2 = k/m$. Multiplying both sides of this equation by 2 *dx/dt*, we have

$$2\frac{dx}{dt}\frac{d^2x}{dt^2} = -\omega^2\left(2x\frac{dx}{dt}\right)$$

Integrating with respect to *t*, we get

$$\left(\frac{dx}{dt}\right)^2 = -\omega^2 x^2 + C\text{ (constant)}.$$

When $x=0, x^2 = \frac{dx}{dt} = v_0$ (given). Therefore,

$$C = v_0^2$$

so that
$$\left(\frac{dx}{dt}\right)^2 = v_0^2 - \omega^2 x^2$$

or
$$\frac{dx}{dt} = \sqrt{v_0^2 - \omega^2 x^2} = \omega\sqrt{\left(\frac{v_0^2}{\omega^2} - x^2\right)}$$

or
$$\frac{dx}{\sqrt{\left(\frac{v_0^2}{\omega^2} - x^2\right)}} = \omega\, dt.$$

Integrating, we get

$$\sin^{-1}\frac{x}{v_0/\omega} = \omega t + \phi\text{ (constant)}$$

or
$$\frac{\omega x}{v_0} = \sin(\omega t + \phi).$$

Again, at $t=0$, $x=0$, $\therefore \phi = 0$. Thus

$$\frac{\omega x}{v_0} = \sin \omega t$$

or $$x = \frac{v_0}{\omega} \sin \omega t \,. \qquad \text{...(i)}$$

This is the expression for the position x of the particle as a function of time t.

On differentiating eq. (i), we get dx

$$\frac{dx}{dt} = v_0 \cos \omega t$$

or $$v = v_0 \cos \omega t \,. \qquad \text{...(ii)}$$

This is the expression for the velocity v, as a function of t.

2 . If the displacement of a moving particle at any time is given by $x = a \cos \omega t + b \sin \omega t$, show that the motion is simple harmonic. If $a = 3$, $b = 4$, $\omega = 2$, find the period, maximum velocity and maximum acceleration.

Solution: The displacement is given by

$$x = a \cos \omega t + b \sin \omega t \qquad \text{...(i)}$$

so that $$\frac{dx}{dt} = -\omega a \sin \omega t + \omega b \cos \omega t \qquad \text{...(ii)}$$

and $$\frac{d^2x}{dt^2} = -\omega^2 a \cos \omega t - \omega^2 b \sin \omega t$$

$$= -\omega^2 (a \cos \omega t - b \sin \omega t) \qquad \text{...(iii)}$$

$$= -\omega^2 x$$

The acceleration d^2x/dt^2 is proportional to displacement x and directed opposite to it. Hence the motion is simple harmonic of amplitude $A = \sqrt{(a^2 + b^2)} = \sqrt{(3^2 + 4^2)} = 5$ cm.

The period of motion is

$$\therefore T = \frac{2\pi}{\omega} = \frac{2 \times 3.14 \text{ rad}}{2 \text{ rad/sec}} = 3.14 \text{ sec}.$$

The maximum velocity is

$$\omega A = 2\text{x}5 = 10 \text{ cm/sec}$$

The maximum acceleration is

$$\omega^2 A = (2)^2\text{x } 5 = 20 \text{ cm/sec}^2.$$

3. A body on the end of a spring oscillates with an amplitude of 5 cm at a frequency of 1 Hz. At t=0 the body is at its equilibrium position x=0. Written the equation describing the position of the body, giving the numerical values of A, ω and α in the form $x=A \cos(\omega t+\alpha)$. What will be its velocity at t=8/3 sec.

Solution: The required form of the motion is

$x = A \cos(\omega t + \alpha)$.

Here A=5 cm and $\omega = 2\pi n = 2\pi \times 1 = 2\pi$ rad/sec. Further at t=0, x=0 so that

$$0 = A \cos \alpha$$

or

$$\alpha = \pi/2$$

Hence the equation of motion is

$$x = 5 \cos\left(2\pi t \frac{\pi}{2}\right) \text{cm}$$

or

$$x = 5 \cos\left(2t + \tfrac{1}{2}\right)\pi \text{ cm}.$$

The instantaneous velocity of the body is

$$\frac{dx}{dt} = -10\,\pi \sin\left(2t + \tfrac{1}{2}\right)\pi.$$

At = t = $\frac{8}{3}$ sec, we have

$$\frac{dx}{dt} = -10\,\pi \sin \frac{35}{6}\pi$$

$$= -10\,\pi \sin\left(6\pi - \frac{\pi}{6}\right)$$

$$\left[\because \sin\left(6\pi - \frac{\pi}{6}\right) = -\sin\frac{\pi}{6}\right]$$

$$= 10\,\pi \sin\frac{\pi}{6} \qquad \left[\because \sin\frac{\pi}{6} = \frac{1}{2}\right]$$

$$= 5\,\pi \text{ cm.}$$

4. A particle of mass 10 gm is placed in a potential field given by V=(50x^2+100) erg/gm. Calculate the frequency of oscillation.

Solution: The potential energy of the 10-gm particle is

$$(U = 10\,(50\,x^2+100)\text{ erg.}$$

The force acting upon the particle is given by

$$F = -\frac{dU}{dx} = -1000\,x \text{ dyne.}$$

But $F = m\dfrac{d^2x}{dt^2}$, by Newton's law.

$$\therefore\ m\frac{d^2x}{dt^2} = -1000\,x$$

or
$$\frac{d^2x}{dt^2} = -\frac{1000\,x}{m} = -\frac{1000\,x}{10} = -100\,x$$

or
$$\frac{d^2x}{dt^2} = -\omega^2 x,$$

where $\omega^2 = 100$. Since the acceleration is proportional to the displacement, the particle is executing S. H. M. The frequency of oscillation is

$$n = \frac{\omega}{2\pi} = \frac{10}{2\pi} = 1.58 \text{ sec}^{-1}.$$

5. Distinguish between two oscillations given by

$$y_1 = a \cos \omega t \text{ and } y_2 = A \cos(\omega t + \pi)$$

Solution: The oscillations have an *initial* phase difference of π. At start (t=0), the first particle is at one extreme end of its motion ($y_1=A$) while the second particle is at the opposite end of its motion ($y_2 = -A$).

6. Write the equation of S.H.M. if (i) initial phase is zero, (ii) initial phase is $\pi/2$. The amplitude is 5 cm and period is 8 sec.

Solution: The general equation of S. H. M. is

$$x = a \sin(\omega t+\phi)$$

Here $a = 5$ cm, $\omega = \dfrac{2\pi}{T} = \dfrac{2 \times 3.14 \text{ rad}}{8 \text{ sec}} = 0.785$ rad/sec.

(i) When initial phase ϕ=0, we have

$$x = 5 \sin (0\text{-}785\ t) \text{ cm.}$$

(ii) When initial phase $\phi = \pi/2$, we have

$$x = 5 \sin\left(0.785\,t + \frac{\pi}{2}\right)$$

$$=5 \cos (0\text{-}785t).$$

7. A particle executes S. H. M. of period 31.4 sec and amplitude 5 cm. Calculate its maximum velocity and maximum acceleration.

Solution: The equation of S.H.M. is

$$x = a \sin (\omega t+\phi).$$

The velocity is

$$u = \frac{dx}{dt} = \omega \cos (\omega t+\phi).$$

The maximum value of $\cos(\omega t+\phi)$ is 1. Therefore

$$U_{max} = \omega a.$$

$$\text{Here } \omega = \frac{2\pi}{T} = \frac{2 \times 3.14 \text{ rad}}{31.4 \text{ sec}} = 0.2 \text{ rad/sec and } a = 5 \text{ cm}.$$

$$\therefore \; U_{max} = 0.2 \frac{\text{rad}}{\text{sec}} \times 5 \text{ cm} = 1.0 \text{ cm/sec}.$$

The acceleration is given by

$$f = \frac{d^2x}{dt^2} = \omega^2 a \sin(\omega t + \phi)$$

The maximum value of $\sin(\omega t+\phi)$ is 1. Therefore

$$|f_{max}| = \omega^2 a$$

$$= (0.2 \text{ rad/sec})^2 \, (5 \text{ cm})$$

$$= 0.2 \text{ cm/sec}^2.$$

8. Show that when a particle is moving in S.H.M., its velocity at a distance $\sqrt{3}/2$ of its amplitude from the central position is half its velocity in the central position.

Solution: The displacement of a particle in S.H.M. is

$$x = a \sin(\omega t + \phi).$$

The velocity is

$$u = \frac{dx}{dt} = \omega a \cos(\omega t + \phi).$$

$$= \omega a \sqrt{\left(1 - \frac{x^2}{a^2}\right)} = \omega \sqrt{(a^2 - x^2)}.$$

At x—$\sqrt{3}a/2$, the velocity is

$$u = \omega \sqrt{\left(a^2 - \frac{3a^2}{4}\right)} = \tfrac{1}{2} \omega a,$$

which is half the velocity (ωa) in the central position (x=0).

9. The equation of motion of a particle is $x=2 \sin\left(\frac{\pi t}{2} + \frac{\pi}{4}\right)$ cm.

Find the period and the maximum velocity of the particle,

Solution: The given equation is

$$x = 2 \sin\left(\frac{\pi t}{2} + \frac{\pi}{4}\right) \text{cm.}$$

Comparing it with the general equation of S.K.M.

$$x = a \sin(\omega t + \phi),$$

we have

$$a = 2 \text{ cm}, \quad \omega = \frac{\pi}{2} \text{ rad/sec.}$$

The period is

$$T = \frac{2\pi}{\omega} = \frac{2\pi \text{ rad}}{\frac{\pi}{2} \text{ rad/sec}} = 4 \text{ sec.}$$

The maximum velocity of the particle is

$$U_{max} = \omega a$$

$$= \frac{\pi}{2} \text{ rad/sec} \times 2 \text{ cm}$$

$$= 3.14 \text{ cm/sec.}$$

10. The period of a S. H. M. is 10 sec and amplitude is 10^{-1} m. Write its equation. What are the phase and displacement at a time 5 sec after a passage of the particle through its extreme positive elongation ? What is the maximum velocity?

Solution: The general equation of S. H. M. is

$$x = a \sin(\omega t + \phi)$$

Here $a = 10^{-1}$ m $= 10$ cm and $\omega = \frac{2\pi}{T} = \frac{2 \times \pi}{10 \text{ sec}} = \frac{\pi}{5}$ rad/sec.

$$\therefore\ x = 10 \sin\left(\frac{\pi t}{5} + \phi\right) \text{cm.}$$

If the motion starts from the extreme positive elongation, that is,

$$x = a \sin(\omega t + \phi)$$

then we have

$$10 = 10 \sin \phi$$

or $$\phi = \frac{\pi}{2}$$

The equation of motion is

$$x = 10 \sin\left(\frac{2\pi}{5} + \frac{\pi}{2}\right) \qquad ...(i)$$

Now, at $t = 5$ sec, the displacement is

$$x = 10 \sin\left(\frac{2 \times \pi}{5} + \frac{\pi}{2}\right)$$

$$= 10 \sin \frac{3\pi}{2} = -10 \text{ cm.}$$

The phase at t=5 sec is $\frac{3\pi}{2}$.

The velocity is, differentiating (i), given by dx

$$u = \frac{dx}{dt} = \left(\frac{\pi}{5}\right) 10 \cos\left(\frac{\pi t}{5} + \frac{\pi}{2}\right)$$

$$\therefore \; u_{max} = \left(\frac{\pi}{5}\right) 10 = 2\pi = 6.28 \text{ cm/sec}.$$

11. Two particles 1 and 2 start vibrating together in S.H.M. along the same straight line. If their periods are 40 and 60 sec respectively, find their phase difference (a) after 20 sec from start and (b) when one particle is at the end of its path while the other is at the middle.

Solution: The displacement of a particle in S. H. M. about x=0 is given by

$x = a \sin \omega t$,

where a is amplitude and ωt is the phase of motion. The initial phase (at t=0) has been assumed to be zero. If T be the period, then

$$x = a \sin \frac{2\pi}{T} t$$

(a) For particles 1 and 2, T=40 sec and 60 sec respectively.

Thus

$$x_1 = a \sin \frac{\pi t}{20}$$

and
$$x_2 = a \sin \frac{\pi t}{30}$$

At t=20 sec, we have

$$x_1 = a \sin \pi$$

and
$$x_2 = a \sin \frac{2\pi}{3} = a \sin\left(\pi - \frac{\pi}{3}\right).$$

These two eq. show that the phase difference between particles 1 and 2 is $\pi/3$, *i.e.* 60°.

(b) Particle 1 is at the end of its path and particle 2 is at the middle, *i.e.* $x_1 = \pm a$ and $x_2 = 0$. Thus at this instant, we can write

$$\pm a = a \sin \omega_1 t$$

and
$$0 = a \sin \omega_2 t.$$

These give

$$\omega_1 t = \frac{\pi}{2} \text{ or } \frac{3\pi}{2} \text{ and } \omega_2 t = 0$$

$\therefore$ phase difference

$$\omega_1 t \sim \omega_2 t = \frac{\pi}{2} \text{ or } \frac{3\pi}{2}.$$

12. A particle of mass 5 gm executing S. H. M. has amplitude of 8 cm. If it makes 16 vibrations per second, find its maximum velocity and energy at mean position.

Solution: The displacement of a particle in S. H. M. is

$$x = a \sin(\omega t + \phi),$$

and its velocity

$$u = \frac{dx}{dt} = \omega a \cos(\omega t + \phi)$$

It is maximum when $\cos(\omega t+\phi) = 1$, *i.e.*

$$u_{max} = \omega a$$

$$= 2\pi na \quad \text{where } n \text{ is frequency}$$

$$= 2\times 3.14\times 16\times 8 = 803.8 \text{ cm/sec.}$$

The energy at mean position is entirely kinetic and is given by

$$E = K_{max} = \tfrac{1}{2} m u_{max}^{2}$$

$$= \tfrac{1}{2}\times 5\times (803.8)^2 = 1.6\times 10^6 \text{ erg.}$$

13. A particle in S.H.M. has velocities u_1 and u_2 when its displacements from the mean position x_1 and x_2 respectively. Calculate the period, amplitude and maximum speed of the particle.

Solution: The speed of a particle in S. H. M. at a displacement $x = 0$ is given by

$$u = \omega\sqrt{(a^2 - x^2)},$$

where ω is angular speed and a is amplitude.

Here.
$$u_1 = \omega\sqrt{(a^2 - x_1^{2})} \quad \text{...(ii)}$$

and
$$u_2 = \omega\sqrt{(a^2 - x_2^{2})}$$

$$u_1 - u_2 = \omega^2 (x_2^{2} - x_1^{2})$$

or
$$\omega = \sqrt{\left(\frac{u_1^{2} - u_2^{2}}{x_2^{2} - x_1^{2}}\right)}$$

The period is therefore

$$T = \frac{2\pi}{\omega} = 2\pi\sqrt{\left(\frac{x_2^2 - x_1^2}{u_1^2 - u_2^2}\right)}$$

From eq. (i) and (ii), we have

$$\frac{u_1^2}{u_2^2} = \frac{a^2 - x_1^2}{a^2 - x_2^2}$$

or $$a^2u_1^2 - x_2^2u_1^2 = a^2u_2^2 - x_1^2u_2^2$$

or $$a = \sqrt{\left(\frac{x_2^2u_1^2 - x_1^2u_2^2}{u_1^2 - u_2^2}\right)}$$

The maximum particle velocity in S. H. M. is

$$u_{max} = \omega a$$

$$= \sqrt{\left(\frac{u_1^2 - u_2^2}{x_2^2 - x_1^2}\right)}\sqrt{\left(\frac{x_2^2u_1^2 - x_1^2u_2^2}{u_1^2 - u_2^2}\right)}$$

$$= \sqrt{\left(\frac{x_2^2u_1^2 - x_1^2u_2^2}{x_2^2 - x_1^2}\right)}$$

14. A particle is moving with simple harmonic motion along a straight line. Its velocity, when passing through the points 3 cm and 4 cm from the centre of its path is 16 cm/sec and 12 cm/sec respectively. Find the amplitude and period of motion and the maximum velocity of the particle.

Solution: The displacement of a particle in S. H. M. is

$$x = a\sin(\omega t + \phi),$$

where a is amplitude. Its velocity

$$\frac{dx}{dt} = \omega a\cos(\omega t + \phi)$$

$$= \omega a\sqrt{\left[1-\sin^2(\omega t+\phi)\right]}$$

$$= \omega a\sqrt{\left[1-\left(x^2/a^2\right)\right]}$$

$$= \omega\sqrt{\left[1-\left(a^2-x^2\right)\right]}$$

or $$\left(\frac{dx}{dt}\right)^2 = \omega^2\left(a^2-x^2\right). \qquad ...(i)$$

Now, it is given that

$$\text{when } x = 3 \text{ cm}, \frac{dx}{dt} = 16 \text{ cm/sec},$$

$$\text{when } x = -4 \text{ cm}, \frac{dx}{dt} = 12 \text{ cm/sec}.$$

Substituting in eq. (i), we get

$$256 = \omega^2 (a^2-9) \qquad ...(ii)$$

and $$144 = \omega^2 (a^3—16). \qquad ..(iii)$$

Dividing, we get

$$\frac{16}{9} = \frac{a^2-9}{a^2-16}$$

or $$16 (a^2-16) = 9 (a^2-9)$$

or $$7a^2 = 256 - 81 = 175$$

or $$a^2 = \frac{175}{7} = 25$$

or $$a = 5\text{cm}.$$

Substituting in eq. (iii), we get

$$144 = \omega^2 (25 - 16)$$

$$\therefore \qquad \omega^2 = \frac{144}{9}$$

or
$$\omega = \frac{12}{3} = 4$$

Therefore, the period is

$$T = \frac{2\pi}{\omega} = \frac{2 \times 3.14}{4} = 1.57 \text{ sec.}$$

Maximum velocity of the particle

$$\left(\frac{dx}{dt}\right)_{max} = \omega a$$

$$=4 \times 5 = 20 \text{ cm/sec.}$$

15. A particle executes linear S. H. M. about the point x=0. At time t=0, it has displacement x=2 cm and zero velocity. If the frequency of motion is 0.25/sec, find (i) period, (ii) circular frequency, (iii) amplitude, (iv) maximum speed, (v) displacement and velocity at t=3 sec.

Calculate the time required for the particle to come halfway in towards the centre of motion from its initial position.

Solution: (i) Frequency, n=0.25 sec^{-1}

$$\therefore \text{ Period, } T = \frac{1}{n} = \frac{1}{0.25} = 4 \text{ sec.}$$

(ii) Circular frequency, $\omega = \frac{2\pi}{T} = \frac{2\pi}{4} = \frac{\pi}{2}$ rad/sec.

(iii)
$$x = a \sin (\omega t + \phi)$$

so that
$$u = \frac{dx}{dt} = \omega a \cos (\omega t + \phi)$$

At t =0 x=2 cm and u=0 (given). Therefore from eq. (i) and (ii), we have

$$2 = a \sin \phi. \qquad \text{...(iii)}$$

and
$$0 = \omega a \cos \phi. \qquad \text{...(iv)}$$

From eq. (iv), cos ϕ0 or $\phi = \pi/2$ or $3\pi/2$. Therefore from eq. (iii), we get

$$a = \pm 2 \text{ cm. } (\because \sin \pi/2 = 1; \sin 3\pi/2 = -1)$$

(iv) From eq. (ii), we have

$$u_{max} = \omega a = \frac{\pi}{2} \times 2 = 3.14 \text{ cm/sec.}$$

(v) Putting $t = 3$ sec, $\omega = \pi/2$, a=2 cm and $\phi = \pi/2$ in eq. (i) and (ii), we get

$$x = 2 \sin 2\pi = 0$$

and $$u = 3.14 \cos 2\pi = 3.14 \text{ cm/sec.}$$

Let us now compute the time required for the particle to come halfway in towards the centre of motion from its initial position. Putting a=2 cm, $\omega = \pi/2$ and $\phi = \pi/2$ in eq. (i), we get

$$x = 2 \sin\left(\frac{\pi t}{2} + \frac{\pi}{2}\right) = 2 \cos\frac{\pi t}{2}.$$

At halfway, x= $\frac{a}{2}$ = 1 cm. Hence,

$$1 = 2 \cos \frac{\pi t}{2}$$

or $$\cos \frac{\pi t}{2} = \tfrac{1}{2}$$

or $$\frac{\pi t}{2} = \cos^{-1}\left(\tfrac{1}{2}\right) = \frac{\pi}{3}$$

$\therefore$ $$t = \frac{2}{3} \text{ sec.}$$

16. The equation of motion of an oscillating body is $x = 6 \cos\left(3\pi t + \frac{\pi}{3}\right)$ metre. What is the period, frequency and phase

constant of the motion ? Also find the displacement, velocity and acceleration at time t=2 sec.

Solution: The given equation

$$x=6 \cos\left(3\pi t+\frac{\pi}{3}\right) \text{metre}$$

represents a S. H. M. Let us compare it with the general equation of S. H. M. in the cosine form which is

$x=a \cos(\omega a+\phi)$

We see that for the given oscillating body

a = 6 metre, ω =3 π and ϕ= π /3.

$\therefore$ period, $T= \frac{2\pi}{\omega}=\frac{2\pi}{3}=\frac{2}{3}$ second,

frequency, $n = \frac{1}{T}=\frac{3}{2}$ vib/sec,

and phase constant, $\phi = \frac{\pi}{3}$ rad=60°.

Now, the displacement is

$$x = 6 \cos\left(3\pi t + \frac{\pi}{3}\right) \text{ metre.}$$

At t = 2 sec, $x = 6 \cos\left(6\pi t + \frac{\pi}{3}\right)$

$$= 6 \cos \frac{\pi}{3} = 6 \cos 60° = 3 \text{ metre.}$$

The velocity is $u = \frac{dx}{dt} = -18\pi \sin\left(3\pi t + \frac{\pi}{3}\right)$ metre/sec.

At t=2 sec, $u = -18\pi \sin\left(6\pi t + \frac{\pi}{3}\right)$

$$= -18\pi \sin\frac{\pi}{3} = -18\pi \times \sin 60^0$$

$$= -18\pi \times \frac{\sqrt{3}}{2} = — 49 \text{ metre/sec.}$$

The acceleration is

$$f = \frac{d^2x}{dt^2} = -54\pi^2 \cos\left(3\pi t + \frac{\pi}{3}\right) \text{metre/sec}^2$$

at t=2 sec, $$f = -54\pi^2 \cos\left(6\pi t + \frac{\pi}{3}\right)$$

$$= -54\pi/3 = -54\pi^2 \cos 60°$$

$$= -54\pi^2 \times \tfrac{1}{2} = -266 \text{ metre/sec}^2.$$

17. A mass of 250 gm hangs from a vertical spring of elastic constant 2 N/m. Calculate the frequency of small vertical oscillations, assuming that the damping is small.

Solution: m=250 gm=0.25 kg and k=2 newton/metre.

$$\therefore n = \frac{1}{2\pi}\sqrt{\left(\frac{k}{m}\right)}$$

$$= \frac{1}{2 \times 3.14}\sqrt{\left(\frac{2}{0.25}\right)} = 0.45 \text{ sec}^{-1}.$$

18. A spring is hung vertically and loaded with a mass of 400 gm and made to oscillate. Calculate (i) the time-period and frequency of oscillation. When the spring is loaded with 100 gm, it extends by 5 cm.

Solution: If an applied force F_{app} extends a spring by x, then by Hooke's law $F_{app} = kx$,

where k is force-constant of the spring. Here F_{app} = 100 gm = 100 x 980 dynes and x= 5 cm.

$$\therefore k = \frac{F_{app}}{x} = \frac{100 \times 980}{5} = 1.96 \times 10^4 \text{ dyne/cm.}$$

Now, the period of a mass m hung from a light spring is given

$$T = 2\pi\sqrt{\left(\frac{m}{k}\right)}$$

'Here m = 400 gin and $k = 1.96 \times 10^4$ dyne/cm.

$$\therefore \quad T = 2\pi\sqrt{\left(\frac{400}{1.96\times10^4}\right)}$$

$$=2 \times 3.14\times\sqrt{\left(\frac{1}{49}\right)} =0.9 \text{ sec.}$$

The frequency is $n = \frac{1}{T} = \frac{1}{0.9 \text{ sec}} = 1.1 \text{ sec}^{-1}$.

19. A mass M of 2000 gm hangs from a vertical spring. When a mass m of 600 gm is gently added to 2000 gm, the spring is further stretched by 4 cm. Now m is removed and M is set in oscillation. Calculate the period.

Solution: Suppose the spring is of force-constant k and is stretched by x cm due to the hanging mass M. Then

$$F_{app} = Mg = kx$$

or $$2000\ g = kx. \qquad ...(i)$$

When the mass m=600 gm is added, the spring is further stretched by 4 cm. That is

$$(M+m)\ g=k\ (x+4)$$

or $$(2000+600)\ g=k\ (x+4). \qquad ...(ii)$$

Solving eq. (i) and (it), we get

$$k = 150\ g$$

$$= 150 \times 980 = 147000 \text{ dyne/cm.}$$

Now, the period of oscillation of M is

$$T = 2\pi\sqrt{\frac{M}{k}}$$

$$= 2\times 3.14\times \times\sqrt{\left(\frac{2000}{147000}\right)} = 0.73 \text{ sec.}$$

20. A body of mass 0.1 kg hanging from a spring is oscillating with a period of 0.2 second and an amplitude of 1 metre. Find the maximum value of the force acting on the body and the value of the force-constant of the spring.

Solution: The equation of S. H. M. of a body of mass m is

$$x = a\ (\omega t+\phi)$$

where a is the amplitude and $\omega = \sqrt{\left(\frac{k}{m}\right)}$. The time-period is

$$T = \frac{2\pi}{\omega} = 2\pi\sqrt{\left(\frac{m}{k}\right)}.$$

Here T=0.2 sec.

$$\therefore \quad \omega = \sqrt{\left(\frac{k}{m}\right)} = 10\pi. \qquad \text{...(ii)}$$

The acceleration of the body is, differentiating (i) twice

$$f = \frac{d^2x}{dt^2} = -\omega^2 a \sin(\omega t+\phi)$$

Tlieit $f_{max} = \omega^2 a$.

Here $\omega = 10\pi$ and a=1 metre (given).

$$\therefore f_{max} = (10\pi)^2 \times 1 = 986$$

and max. force=mass × max. acceleration

= (0.1 kg) × (986 metre/sec^2)

=98.6 newton.

Now, from eq. (ii), we have

$$\frac{k}{m} = (10\pi)^2$$

$$\therefore \quad k = 100\,\pi^2 m$$

$$= 100 \times (3.14)^2 \times 0.1$$

$$= 986 \text{ newton/metre.}$$

21. The potential energy of a harmonic oscillator in its rest position is 5 joule and the average kinetic energy is 2 joule. What is the total energy ? If the amplitude is 1 metre, what is the force-constant? If the mass is 2 kg, what is the period ?

Solution: The potential energy in rest position is 5 joule. The average kinetic energy (2 joule) is half the energy of oscillation so that the energy of oscillation is 4 joule. Hence the total energy

$$= 5+4=9 \text{ joule.}$$

At the maximum displacement *a*, the gain in potential energy is $\frac{1}{2}ka^2$, where *k* is force-constant. Thus

$$\tfrac{1}{2}ka^2 = 4 \text{ joule.}$$

Here a = 1 metre.

$$\therefore \; k = 8/a^2 = 8 \text{ newton/metre.}$$

The period of oscillation is

$$T = 2\pi\sqrt{\left(\frac{m}{k}\right)} = 2\pi\sqrt{\left(\frac{2\,\text{kg}}{8\,\text{nt/m}}\right)} = 3.14 \text{ sec.}$$

22. A simple harmonic oscillator of period 5 sec has 5 joule potential energy when its displacement is 2 cm. (Energy in the rest position may be taken as zero). Calculate (i) force-constant, (ii) average kinetic energy if the amplitude is 4 cm, (iii) kinetic and potential energies when the displacement is 1 cm,

(iv) oscillation frequency if mass is reduced to $\frac{1}{100}$ th of its initial value.

Solution: (i) The gain in potential energy when the displacement is x is $\frac{1}{2}kx^2$, where k is force-constant. Here at x=2 cm=0.02 metre, the gain in potential energy is 5 joule. Thus

$$\tfrac{1}{2}k\,(002)^2 = 5$$

or

$$k = 2.5 \times 10^4 \text{ newton/metre.}$$

(ii) The average kinetic energy is

$$K_{av} = \tfrac{1}{4}ka^2.$$

Here a=4 cm=4 × 10^{-2} metre.

$$\therefore K_{av} = \tfrac{1}{4} \times (2.5 \times 10^4 \text{ nt/m}) \times (4 \times 10^{-2} \text{ m})^2$$

$$=10 \text{ joule.}$$

(iii) The kinetic energy when the displacement is x is

$$K = \tfrac{1}{2}k\,(a^2-x^2).$$

Here a=4 × 10^{-2} metre and. x=l × 10^{-2} metre.

$$\therefore K = \tfrac{1}{2} \times (2.5 \times 10^4) \times (15 \times 10^{-4})$$

$$= 18.75 \text{ joule.}$$

The potential energy is

$$U = \tfrac{1}{2}kx^2$$

$$= \tfrac{1}{2}(2.5 \times 10^4 \text{ nt/m})\,(1 \times 10^{-2} \text{ m})^2$$

$$=1.25 \text{ joule.}$$

(iv) The frequency is inversely proportional to the square-root of mass. The initial frequency is

$$n_i = \frac{1}{5 \text{ sec}} = 0.2 \text{ sec}^{-1}$$

If m_4 and m_f be the initial and the final masses, the final frequency is given by

$$\frac{n_i}{n_i} = \sqrt{\left(\frac{m_i}{m_f}\right)}.$$

Here $m_f = \dfrac{1}{100} m_i$

$$\therefore \frac{n_f}{0.2} = \sqrt{(100)} = 10$$

or $$n_f = 0.2 \times 10 = 2 \text{ sec}^{-1}.$$

23. A body of mass 0.5 kg is projected with an energy 4 joules. If a restoring force of 2 nt/m acts on it, what maximum displacement will the body acquire? What will be its periodic time?

Solution: At the maximum displacement *a* (say), the energy would be entirely potential energy. That is,

$$\tfrac{1}{2} k\, a^2 = 4 \text{ joule.}$$

Here k = 2 nt/m.

$$a = \sqrt{\left(\frac{2 \times 4\text{joule}}{2\,\text{nt/m}}\right)} = 2 \text{ metre}$$

The period is

$$a = 2\pi \sqrt{\left(\frac{m}{k}\right)}$$

$$= 2\pi \sqrt{\left(\frac{0.5\,\text{kg}}{2\,\text{nt/m}}\right)} = \pi \text{ sec.}$$

24. The amplitude of harmonic oscillations of a point-mass of 10 gm is 5 cm and the energy of oscillation is 3.1×10^{-5} joule. Write the equation of motion if the initial phase is 60°.

Solution: The energy of oscillation is given by

$$E = 2\pi^2 m\, n^2\, a^2 = \tfrac{1}{2}(2\pi n)^2\, ma^2$$

$$= \tfrac{1}{2}\,\omega^2\, ma^2. \qquad (\because\ \omega = 2\pi n)$$

$$\therefore\ \omega = \frac{1}{a}\sqrt{\frac{2E}{m}}.$$

Here E=3.1 × 10^{-5} joule=310 erg. ($\because$ 1 joule=10^7 erg), a=5 cm and m = 10 gm.

$$\therefore\ \omega = \tfrac{1}{5}\sqrt{\frac{2\times 310}{10}} = \frac{\sqrt{62}}{5} = 1.57 \text{ rad/sec}$$

Now, the equation of motion is

$$x = a\sin(\omega t + \phi)$$

Putting a=5 cm, ω =1.57 rad/sec, ϕ =60°= π /3 rad, we get

$$x = 5\sin\left(1.57t + \frac{\pi}{3}\right) \text{ cm.}$$

25. A mass M is controlled by two massless springs between two rigid supports as shown. If the mass is 50 gm and the force constants of springs are 3000 and 2000 dyne/cm, find (i) frequency of oscillations of the mass, (ii) energy of vibration of amplitude 0.4 cm, (iii) velocity of mass when passing through the mean position.

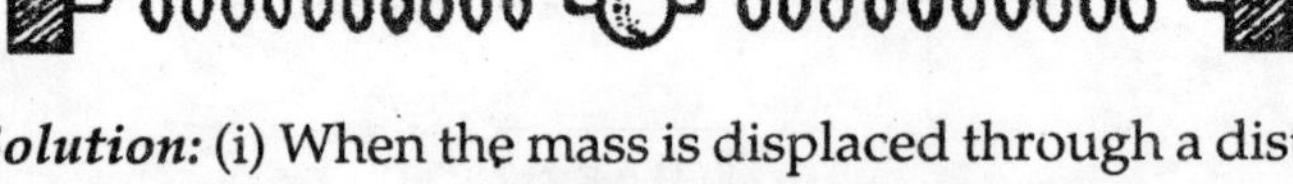

Solution: (i) When the mass is displaced through a distance x, the restoring forces developed in the springs are

$$F = -k_1 x \text{ and } F_2 = -k_2 x,$$

and act in the same direction. The total restoring force is

$$F = -(k_1 + k_2)\, x,$$

so that the force-constant of the system is

$$k = k_1 + k_2 = 3000 + 2000 = 5000 \text{ dyne/cm.}$$

Therefore, the frequency of oscillation is

$$n = \frac{1}{2\pi}\sqrt{\left(\frac{k}{m}\right)} = \frac{1}{2\pi}\sqrt{\left(\frac{5000}{50}\right)} = \frac{5}{\pi}$$

(ii) The energy of oscillation is the same as the maximum potential energy. Thus

$$E = U_{max} = \frac{1}{2}ka^2, \qquad \text{where } a \text{ is amplitude}$$

$$\therefore\ E = \frac{1}{2} \times 5000 \times (0.4)^2 = 400 \text{ erg.}$$

(iii) The velocity, and hence the kinetic energy, is maximum when the mass passes through its mean position. Again, the energy of oscillation is same as the maximum kinetic energy.

$$\therefore\quad E = K_{max} = \frac{1}{2}\ m\ (U_{max})^2 = 400$$

$$\therefore\quad U_{max} = \sqrt{\left(\frac{800}{m}\right)} = \sqrt{\left(\frac{800}{50}\right)} = \pm 4 \text{cm/sec.}$$

26. A rifle ballet weighing 10 gm moving with a velocity of 16000 cm/sec. strikes and embeds itself in a 790-gm block which rests on a horizontal frictionless surface and is attached to a spring of force constant 8×10^4 dyne/cm. Compute the amplitude of the resulting simple harmonic oscillations of the block. What happens to the kinetic energy of the bullet?

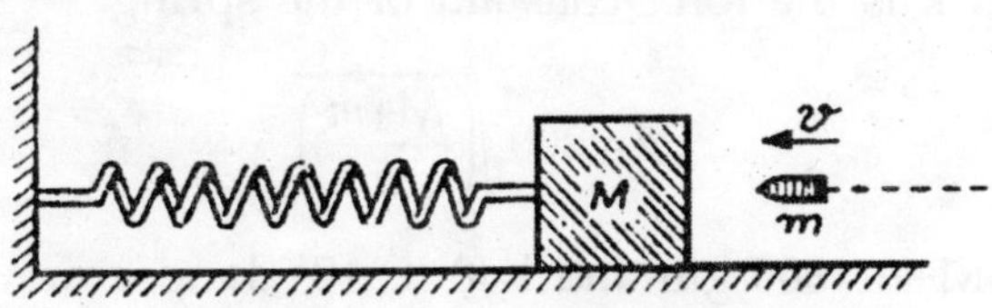

Solution: Let v' be the velocity acquired by the block M when the bullet m strikes it and comes to rest in it. By conservation of momentum, we have

$$(M+m)\, v' = mv$$

$$800 x v' = 10 \times 16000$$

$$\therefore \quad v' = \frac{10 \times 16000}{800} = 200 \text{ cm/sec.}$$

The block is set in oscillation about its mean position. If a be the amplitude of motion, the equation of motion may be written as

$$x = a \sin \omega t\,.$$

The velocity is

$$\frac{dx}{dt} = \omega a \cos \omega t.$$

In the mean position, the velocity is a maximum and equal to *wa*. This we have obtained as 200 cm/sec. Thus

$$\omega a = 200$$

or

$$a = \frac{200}{\omega}$$

If T be the period, then $T = \dfrac{2\pi}{\omega}$, *i.e.* $\omega = \dfrac{2\pi}{T}$

$$\therefore \quad a = \frac{200\,T}{2\pi}$$

But $T = 2\pi\sqrt{\left(\dfrac{M+m}{k}\right)}$, where M+m is the mass of the block+ bullet and k is the force-constant of the spring.

$$\therefore a = 200\sqrt{\left(\frac{M+m}{k}\right)}$$

Here M+m=800 gm and k=8 $\times$ 10^4 dyne/cm.

$$\therefore a = 200 \times \sqrt{\left(\frac{800}{8 \times 10^4}\right)} = 20 \text{ cm.}$$

The kinetic energy of the moving bullet was

$$K = \tfrac{1}{2} mv^2$$

$$= \tfrac{1}{2} 10 \times (16000)^2 = 128 \times 10^7 \text{ erg.}$$

The maximum (potential) energy stored in the spring is

$$U = \tfrac{1}{2} ka^2$$

$$= \tfrac{1}{2} \times (8 \times 10^4) \times (20)^2 = 1.6 \times 10^7 \text{ erg}$$

$\therefore$ percentage of energy stored in the spring

$$= \frac{1.6 \times 10^7}{128 \times 10^7} \times 100 = 1.25 \text{ per cent}$$

Thus a very small fraction of the kinetic energy of the bullet is stored in the spring and the rest appears as heat.

27. A block rests on a horizontal surface which is executing S. H. M. in a horizontal plane at the rate of 2 oscillations/sec. The coefficient of static friction μ between block and plane is 0.5. How large can the amplitude be without slipping between block and surface?

Solution: Let m be the mass of the block and a the largest possible amplitude without slipping between block and surface. Then the maximum horizontal force acting on the block must be equal to the frictional force. The maximum horizontal force on the block is $m\omega^2 a$, because $\omega^2 a$ is the maximum acceleration where ω is the angular frequency. The frictional force is μmg. Thus

$$m\omega^2 a = \mu mg$$

or

$$a = \frac{\mu g}{\omega^2}$$

But $\omega = 2\pi n$, where n is the frequency.

$$\therefore\ a = \frac{\mu g}{4\pi^2 n^2}$$

$$= \frac{0.5 \times 980}{4 \times (3.14)^2 \times (2)^2} = 3.1 \text{ cm.}$$

28. A simple pendulum of length 2 metre and amplitude 0.03 metre has energy 0.10 joule. What will be its energy (i) if its amplitude is changed to 0.06 metre and length remains unchanged, (ii) amplitude remains 0.03 metre but length is changed to 1 metre.

Solution: The energy of oscillation of amplitude a is

$$E = \tfrac{1}{2} m\omega^2 a^2$$

In case of simple pendulum, $\omega = \frac{2\pi}{T} = \sqrt{\left(\frac{g}{l}\right)}$.

$$\therefore \quad E = \frac{mga^2}{2l}$$

Originally, l=2 metre, a=0.03 metre, E=0.10 joule.

$$\therefore \; 0.10 = \frac{mg \times (0.03)^2}{2 \times 2}. \qquad \text{...(i)}$$

(i) $$E = \frac{mg \times (0.06)^2}{2 \times 2}. \qquad \text{...(ii)}$$

Dividing eq. (ii) by eq. (i), we get

$$E = \frac{(0.06)^2}{(0.03)^2} \times 0.10 = 0.40 \text{ joule.}$$

(ii) $$E = \frac{mg \times (0.03)^2}{2 \times 1}. \qquad \text{...(iii)}$$

Dividing eq. (iii) by eq. (i), we get

$$E = \tfrac{3}{1} \times 0.10 = 0.20 \text{ joule.}$$

29. One end of a horizontal spring of force constant 6.0 newton/metre is tied to a fixed wall and the other end to a

solid cylinder which can roll without slipping on the horizontal ground. The cylinder is polled a distance of 0.1 metre and released. Calculate the kinetic energies of the cylinder when passing through the equilibrium position.

Solution: When the cylinder is pulled 0.1 metre, the spring is stretched and has a potential energy given by

$$U=\tfrac{1}{2}kx^2=\tfrac{1}{2}\times 6.0\times(0.1)^2=0.03 \text{ joule.}$$

Whole of this energy is converted into the kinetic energy of rotation and the kinetic energy of translation when the cylinder is passing through the equilibrium position. That is

$$K_{rot}+K_{trans}=0\text{-}03 \text{ joule.}$$

$$K_{rot}=\tfrac{1}{2}I\omega^2$$

$$=\tfrac{1}{2}\left(\tfrac{1}{2}mr^2\right)\frac{v^2}{r^2}=\tfrac{1}{4}mv^2,$$

and $$K_{trans}=\tfrac{1}{2}mv^2$$

$$\therefore\ \tfrac{1}{4}mv^2+\tfrac{1}{2}mv^2=0.03 \text{ joule.}$$

or $$mv^2=0.03\times\tfrac{4}{3}=0.04 \text{ joule.}$$

$$\therefore\ K_{rot}=\tfrac{1}{4}mv^2=\tfrac{1}{4}(0.04)=0.01 \text{ joule.}$$

and $$K_{trans}=\tfrac{1}{2}mv^2=\tfrac{1}{2}(0.04)=0.02 \text{ joule.}$$

3

Addition of Simple Harmonic Motions

Addition of two S. H. M.'s having Equal Periods in a Straight Line: When a particle is acted upon simultaneously by simple harmonic motions in the same straight line, its displacement at any instant is the algebraic sum of the individual displacements for the two separate motions.

Let the two simple harmonic motions of the same period be represented by

$$x_1 = a_1 \sin(\omega t + \phi_1) \qquad \text{...(i)}$$

and

$$x_2 = a_2 \sin(\omega t + \phi_2), \qquad \text{...(ii)}$$

where a_1 and a_2 are the amplitudes and ϕ_1 and ϕ_2 the initial phases of the two motions. The period of each is $2\pi/\omega$. The resultant displacement at any instant t is then given by $x = x_1 + x_2$

$$= a_1 \sin(\omega t + \phi_1) + a_2 \sin(\omega t + \phi_2)$$

$$= a_1(\sin\omega t \cos\phi_1 + \cos\omega t \sin\phi_1) + a_2(\sin\omega t \cos\phi_2 + \cos\omega t \sin\phi_2)$$

$$= \sin\omega t\,(a_1 \cos\phi_1 + a_2 \cos\phi_2) + \cos\omega t\,(a_1 \sin\phi_1 + a_2 \cos\phi_2).$$

Let us make a change in constants a_1, a_2, ϕ_1 and ϕ_2 by putting

$$a_1 \cos\phi_1 + a_2\ cas\,\phi_2 = r\ cos\ \theta, \quad \text{...(iii)}$$

and $$a_1 \sin\phi_1 + a_2 \sin\phi_2 — r \sin\theta, \quad \text{... (iv)}$$

where r and θ are new constants. This gives

$$x = \sin\omega t\ (r\cos\theta) + \cos\omega t\ (r\sin\theta)$$

or $$x = r\sin(\omega t + \theta).$$

This equation is of the same type as eq. (i) and (ii) which represent simple harmonic motions. Hence the resultant motion of the particle is also simple harmonic with the same period $(2\pi/\omega)$ as the individual motions, but with a different amplitude r and a different initial phase θ.

To obtain r, let us square and add eq. (iii) and (iv), when we get

$$a_1^2 + a_2^2 + 2a_1a_2(\cos\phi_1\cos\phi_2 + \sin\phi_1\sin\phi_2) = r^2$$

or $$a_1^2 + a_2^2 + 2a_1a_2\cos(\phi_1 - \phi_2) = r^2.$$

Therefore, the resultant amplitude r is given by

$$\boxed{r = \sqrt{\left[a_1^2 + a_2^2 + 2a_1a_2\cos(\phi_1 - \phi_1\right]}.} \quad \text{...(v)}$$

To obtain the resultant initial phase θ, let us divide equation (iv) by equation (iii), when we get

$$\boxed{\tan\theta = \frac{a_1\sin\phi_1 + a_2\sin\phi_2}{a_1\cos\phi_1 + a_2\cos\phi_2}.} \quad \text{...(vi)}$$

There are two special cases :

Maximum Amplitude: When the phase difference between two individual motions is zero, or any integral multiple of 2π, *i.e.*

$$\phi_1 + \phi_2 = 2n\pi,$$

where n= 0, 1, 2,.........., eq. (v) then gives

$$r = \sqrt{\left[a_1^2 + a_2^2 + 2a_1a_2\cos(2n\pi)\right]}$$

$$= \sqrt{\left(a_1^2 + a_2^2 + 2a_1a_2\right)} \qquad [\therefore \cos 2n\pi = 1]$$

$$= \sqrt{\left[(a_1 + a_2)^2\right]}$$

$$= (a_1 + a_2).$$

Thus the resultant amplitude is equal to the sum of the amplitudes of the individual motions.

Minimum Amplitude: When the phase difference between the two individual motions is an odd multiple of π, *i.e.*

$$\phi_1 - \phi_2 = (2n+1)\,\pi,$$

where n=0, 1, 2,........, eq. (v) then gives

$$r = \sqrt{\left[a_1^2 + a_2^2 + 2a_1a_2\cos(2n+1)\pi\right]}$$

$$= \sqrt{\left(a_1^2 + a_2^2 - 2a_1a_2\right)} \qquad [\because \cos(2n+1)\pi = -1]$$

$$= \sqrt{(a_1 - a_2)^2}$$

$$= (a_1 - a_2).$$

Thus the resultant amplitude is equal to the difference of the amplitude of the individual motions.

In this special case, if $a_1 = a_2$, then

$$r = a_1 - a_2 = 0.$$

Thus if a particle is acted upon simultaneously by two simple harmonic motions of equal amplitudes and periods but opposite in phase in the same straight line, it will remain at rest.

Addition of Three Collinear S. H. M.

The resultant instantaneous displacement of a particle under three collinear simple harmonic motions is obtained by their algebraic sum, that is,

$$x = a \sin \omega t + b \sin \omega t + c \sin(\omega t + \phi)$$

$$= \sin \omega t\,(a + b + c \cos\phi) + \cos \omega t\,(c \sin \phi). \quad \text{...(i)}$$

Let us make a change in constants a, b, c and ϕ by putting

$$a + b + c \cos \phi = r \cos \theta \quad \text{.....(ii)}$$

and

$$c \sin\phi = r \sin \theta, \quad \text{...(iii)}$$

where r and θ are new constants. Then we have

$$x = \sin \omega t\,(r \cos \theta) + \cos \omega t\,(r \sin \theta)$$

or

$$x = r \sin(\omega t + \theta).$$

Thus the resultant motion is simple harmonic. The resultant amplitude r is, by squaring and adding eq. (ii) and (iii), given by

$$r^2 = (a + b + c \cos \phi)^2 + (c \sin \phi)^2$$

$$= a^2 + b^2 + c^2 + 2ab + 2(a+b)\, c \cos \phi$$

or

$$r = \sqrt{\left[a^2 + b^2 + c^2 - +2ab + 2\,(a+b)\, c \cos \phi.\right]}$$

The resultant phase is, by dividing eq. (iii) by (ii), given by

$$\tan\theta = \frac{c \sin\phi}{a + b + c \cos\phi}.$$

Addition of two S. H. M's having slightly Different Periods

Let us consider two simple harmonic motions of equal amplitudes but slightly different periods (or frequencies), given by

$$x_1 = a \sin \omega_1 t$$

and $\quad x_2 = a \sin \omega_1 t.$

The resultant instantaneous displacement is given by

$$x = x_1 + x_2$$

$$= (a \sin \omega_1 t + a \sin \omega_2 t)$$

$$= 2a \sin\left(\frac{\omega_1 + \omega_2}{2}\right)t \cos\left(\frac{\omega_1 - \omega_2}{2}\right)t$$

$$= \left\{2a \cos\left(\frac{\omega_1 - \omega_2}{2}\right)t\right\} \sin\left(\frac{\omega_1 + \omega_2}{2}\right)t$$

$$= r \sin\left(\frac{\omega_1 + \omega_2}{2}\right)t.$$

This represents a periodic motion (not simple harmonic) of amplitude

$$r = 2a\cos\left(\frac{\omega_1 - \omega_2}{2}\right)t$$

In terms of frequencies ($\omega_1 = 2\pi n_1$ and $\omega_2 = 2\pi n_2$), we have

$$r = 2a\cos 2\pi\left(\frac{n_1 - n_2}{2}\right)t.$$

This shows that the resultant amplitude of motion varies periodically between $\pm 2a$ and zero under the influence of the cosine function.

r will be maximum when

$$\cos 2\pi\left(\frac{n_1 - n_2}{2}\right)t = \pm 1$$

or $\quad 2\pi\left(\frac{n_1 - n_2}{2}\right)t = k\pi$, where k=0, 1, 2, 3, ...

or $$t = \frac{k}{n_1 - n_2}$$

$$= \frac{1}{n_1 - n_2}, \frac{3}{n_1 - n_2}, \frac{5}{n_1 - n_2}, \ldots\ldots$$

Thus the interval between two consecutive maxima is $\frac{1}{n_1 - n_2}$ Again, r will be minimum when

$$\cos 2\pi \left(\frac{n_1 - n_2}{2}\right) t = 0$$

or $$2\pi \left(\frac{n_1 - n_2}{2}\right) t = \left(k + \tfrac{1}{2}\right)\pi, \text{ where } k = 0, 1, 2, 3, \ldots$$

or $$t = \frac{k + \frac{1}{2}}{n_1 - n_2}$$

$$= \frac{1}{2(n_1 - n_2)}, \frac{3}{2(n_1 - n_2)}, \frac{5}{2(n_1 - n_2)}, \ldots\ldots$$

Thus the interval between two consecutive minima is also $\frac{1}{n_1 - n_2}$

Lissajous' Figures

When a particle has superimposed upon it two mutually perpendicular simple harmonic motions simultaneously, the resultant path of the particle is known as a "Lissajous' figure". The form of the figure depends upon the ratio of the frequencies (or periods), the individual amplitudes and the relative phase of the two component motions.

For example, if the two motions are of equal frequencies, the Lissajous' figure is a straight line; ellipse or circle depending upon the amplitudes and phase difference of the motions. For other frequency-ratios other forms of figures are obtained.

Frequencies (or Periods) Exactly Equal: Let us consider the case when a particle has impressed on it a S. H. M. of amplitude a along the x-axis given by

$$x = a \sin (\omega t + \phi) \qquad \ldots(i)$$

and at the same time a S. H. M. of same frequency but different amplitude b along the y-axis given by

$$y = b \sin \omega t, \quad \text{...(ii)}$$

where ^ is the phase difference.

The equation to the curve of resultant motion can be obtained by eliminating t from cq. (i) and (ii). Expanding eq. (i), we get

$$\frac{x}{a} = \sin \omega t \cos \phi + \cos \omega t \sin \phi.$$

But, from eq. (ii), $\sin \omega t = y/b$ and $\cos \omega t = \sqrt{(1 - \sin^2 \omega t)} = \sqrt{[(1 - y^2/b^2)]}$.

$$\therefore \quad \frac{x}{a} = \frac{y}{b} \cos \phi + \sqrt{\left(1 - \frac{y^2}{b^2}\right)} \sin \phi$$

or
$$\frac{x}{a} - \frac{y}{b} \cos \phi = \sqrt{\left(1 - \frac{y^2}{b^2}\right)} \sin \phi$$

Squaring:
$$\frac{x^2}{a^2} + \frac{y^2}{b^2} \cos^2 \phi - \frac{2yx}{ab} \cos \phi = \left(1 - \frac{y^2}{b^2}\right) \sin^2 \phi$$

or
$$\frac{x^2}{a^2} + \frac{y^2}{b^2} - \frac{2yx}{ab} \cos \phi = \sin^2 \phi \quad \text{..(iii)}$$

This represents an ellipse, contained within a rectangle of sides $2a$ and $2b$, but of an eccentricity and at an inclination depending on the phase difference ϕ, and on the individual amplitudes. Thus, in general, the Lissajous' figure (path of the particle) is an ellipse. There are, however, a number of special cases :

(a) If $\phi = 0°$, eq. (iii) becomes :

$$\frac{x^2}{a^2} + \frac{y^2}{b^2} - \frac{2yx}{ab} = 0 \quad [\because \cos 0° = 1, \sin 0° = 0]$$

or $$\left(\frac{x}{a}-\frac{y}{b}\right)^2=0.$$

This represents two coincident straight lines coinciding with the diagonal $\left(y=\frac{b}{a}x\right)$ of the rectangle.

(b) If $\phi=\pi/4$, eq. (iii) becomes

$$\frac{x^2}{a^2}+\frac{y^2}{b^2}-\frac{\sqrt{2}yx}{ab}=\frac{1}{2}\;[\because\ \cos 45^\circ=1,\ \sin 45^\circ=1/\sqrt{2}\,]$$

This represents an oblique ellipse.

Lissajous' Figures

(c) If $\phi=\pi/2$, the eq. (iii) becomes

$$\frac{x^2}{a^2}+\frac{y^2}{b^2}=1,$$

which is an ellipse whose axes coincide with the coordinate axes. If in addition $a=b$, the path of the particle becomes a circle, $x^2+y^2=a^2$.

(d) If $\phi=3\pi/4$, then $\sin\phi=1/\sqrt{2}$ and $\cos\phi=-1/\sqrt{2}$. The eq (iii) now becomes

$$\frac{x^2}{a^2}+\frac{y^2}{b^2}+\frac{\sqrt{2}yx}{ab}=\frac{1}{2}.$$

This again represents an oblique ellipse.

(e) If $\phi=\pi$, the equation (iii) becomes

$$\frac{x^2}{a^2}+\frac{y^2}{b^2}+\frac{2yx}{ab}=0 \qquad [\because\ \cos\pi=-1,\ \sin\pi=0]$$

or $$\left(\frac{x}{a}+\frac{y}{b}\right)^2=0.$$

This represents two coincident straight lines coinciding with the other diagonal $\left(y=-\frac{b}{a}x\right)$ of the rectangle.

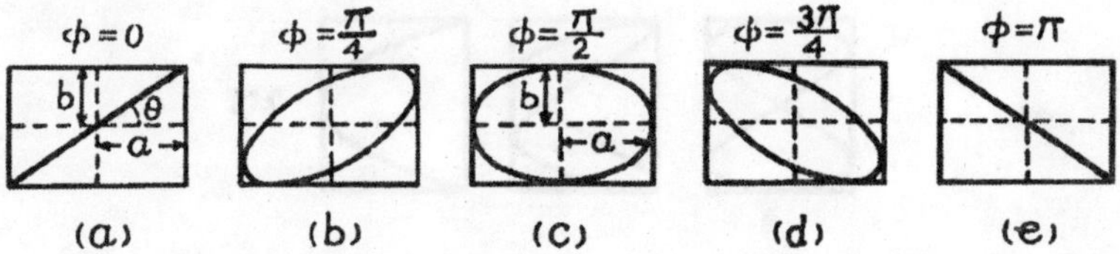

Frequencies Nearly Equal: When the frequencies of the two simple harmonic motions are exactly equal, the elliptic path of the particle remains perfectly steady. If, however, there is a slight difference between the frequencies, the relative phase ϕ of the two component motions slowly changes and accordingly the form of the ellipse changes. Starting with $\phi=0$, the ellipse coincides with the diagonal $\left(y=\frac{b}{a}x\right)$ of the rectangle of sides 2a and *2b*.

As ϕ increases from 0 to $\frac{\pi}{2}$, the ellipse opens out to the form $\frac{x^2}{a^2}+\frac{y^2}{b^2}=1$, afterwards closing up again and ultimately coinciding $\left(y=-\frac{b}{a}x\right)$, as ϕ increases from $\frac{\pi}{2}$ to π. As ϕ changes from a to 2π, the reverse process takes place until the ellipse again coincides with the first diagonal.

One motion gains a complete vibration over the other during the time the phase difference ϕ changes from 0 to 2π and one complete cycle of Lissajous' figures is traced out. Clearly, the frequency of this cycle is equal to the difference of the frequencies of the component motions.

Frequencies Commensurate: If the frequencies of two S.H.M.s' are commensurate, that is, in a definite ratio such as 2:1, 3:1, etc., the particle traces out a curve having two, three, etc. loops. For nearly commensurate frequencies, the curve slowly changes as the phase difference varies. Typical curves for frequency-ratios 2 : 1 and 3:1 are shown in following figure.

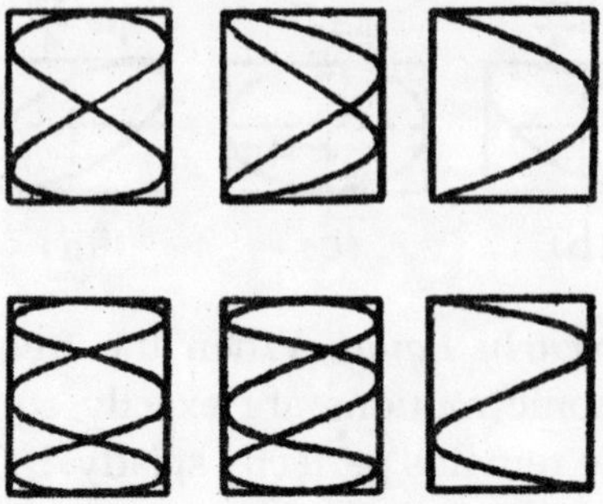

2:1

3:1

When the shape of the curves corresponding to the frequency relations is known, the approximate ratio of the frequencies can be recognised.

Movement of Point in Circle and Clockwise

The S.H.M.s' are

$$x = a_1 \cos \omega t . \qquad \text{...(i)}$$

and

$$y = a_2 \cos\ (\omega t + \phi). \qquad \text{...(ii)}$$

The equation for the path of the particle is obtained by eliminating t between eq. (i) and (ii). From eq. (ii), we have

$$\frac{y}{a_2} = \cos\ \omega t\ \cos\phi - \sin\ \omega t\ \sin\phi .$$

But from eq. (i), $\cos\omega t = x/a_1$ and $\sin\omega t = \sqrt{\left[1-\left(x^2/a_1^{\,2}\right)\right]}$

$$\therefore\ \frac{y}{a_2} = \frac{x}{a_1}\cos\ \phi - \sqrt{\left[1-\left(x^2/a_1^{\,2}\right)\right]}\ \sin^2\phi$$

Squiring :

$$\left(\frac{y}{a_2} - \frac{x}{a_1}\cos\phi\right)^2 = \left(1-\frac{x^2}{a_1^{\,2}}\right)\sin^2\phi$$

Solving :

$$\frac{x^2}{a_1^{\,2}} + \frac{y^2}{a_2^{\,2}} - \frac{2xy}{a_1 a_2}\cos\phi = \sin^2\phi$$

This represents an elliptical path.

If phase difference $\phi = \pi/2$ and $a_1 = a_2 = a$ (say), then

$$x^2 + y^2 = a^2,$$

which represents a circle. Thus the path will be a circle when the two motions have same amplitude and a phase difference of $\pi/2$.

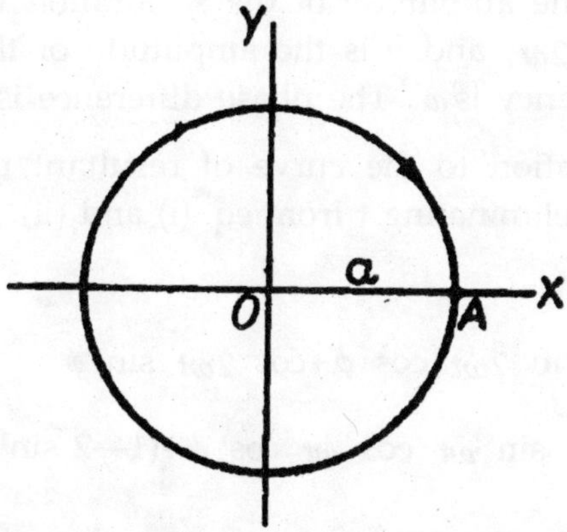

In this case, the x- and y-motions are

$$x = a \cos \omega t$$

$$y = a \cos\left(\omega t + \frac{\pi}{2}\right) = -a \sin \omega t.$$

At the point A, $x — a$, so that $\cos \omega t = 1$. The velocity of the particle at A will be parallel to y- axis, *i.e.*

$$UA = \frac{dy}{dt} = -\omega a \cos \omega t$$

$$= -\omega a \qquad [\because \cos \omega t = 1 \text{ at } A]$$

Since velocity is negative, the particle is moving downward at A, i.e., the circular path is clockwise.

Lissajous' Figures from two Rectangular S.H.M.'s in Frequency Ratio 2:1

Let a particle be subjected to two mutually perpendicular

simple harmonic vibrations along the x- and y-axes and having frequencies in the ratio 2:1. They may be represented by

$$x = a \sin (2\,\omega t+\phi) \qquad \text{...(i)}$$

and

$$y = b \sin \omega t , \qquad \text{...(ii)}$$

where a is the amplitude of the x-vibration whose angular frequency is 2ω, and b is the amplitude of the y-vibration, whose frequency is ω. The phase difference is ϕ.

The equation to the curve of resultant motion can be obtained by eliminating t from eq. (i) and (ii). Expanding cq. (i) we get

$$\frac{x}{a} = \sin\ 2\omega t\ \cos\ \phi + \cos\ 2\omega t\ \sin\ \phi$$

$$= 2 \sin\ \omega t\ \cos\ \omega t\ \cos\ \phi + (1 - 2 \sin^1\ \omega t)\ \sin\ \phi .$$

But from eq. (ii), $\sin\omega t = y/b$ and $\cos\ \omega t = \sqrt{(1-\sin^2 \omega t)} =$

$$\left(1-\frac{y^2}{b^2}\right)$$

$$\therefore \quad \frac{x}{a} = \frac{2y}{b}\sqrt{\left(1-\frac{y^2}{b^2}\right)}\cos\phi + \left(1-\frac{2y^2}{b^2}\right)\sin\phi$$

or

$$\frac{x}{a} - \left(1-\frac{2y^2}{b^2}\right)\sin\ \phi = \frac{2y}{b}\sqrt{\left(1-\frac{y^2}{b^2}\right)}\cos\ \phi$$

Squarring:

$$\frac{x^2}{a^2} + \left(1-\frac{2y^2}{b^2}\right)^2 \sin^2\phi - 2\frac{x}{a}\left(1-\frac{2y^2}{b^2}\right)\sin\ \phi$$

$$= \frac{4y^2}{b^2}\left(1-\frac{y^2}{b^2}\right)\sin\ \phi$$

or $$\frac{x^2}{a^2}+\sin^2\phi-\frac{4y^4}{b^4}\sin^2\phi-\frac{4y^2}{b^2}\sin^2\phi-\frac{2x}{a}\sin^2\phi+\frac{4xy^2}{ab^2}\sin^2\phi$$

$$=\frac{4y^2}{b^2}\cos^2\phi-\frac{4y^4}{b^4}\cos^2\phi$$

or $$\frac{x^2}{a^2}+\sin^2\phi-\frac{2x}{a}\sin^2\phi+\frac{4y^4}{b^4}(\sin^2\phi+\cos^2\phi)$$

$$-\frac{4y^4}{b^4}(\sin^2\phi+\cos^2\phi)+\frac{4xy^2}{ab^2}\sin^2\phi=0$$

or $$\left(\frac{x}{a}-\sin\phi\right)^2+\frac{4y^4}{b^4}-\frac{4y^2}{b^2}+\frac{4xy^2}{ab^2}\sin\phi=0$$

or $$\left(\frac{x}{a}-\sin\phi\right)^2+\frac{4y^2}{b^2}\left(\frac{x^2}{a^2}-1+\frac{x}{a}\sin\phi\right)=0. \quad \text{...(iii)}$$

This is the general equation for a curve having two loops, which is the general path of the particle. Let us consider special cases:

(a) When $\phi=0$, $\sin\phi=0$ and the eq. (iii) reduces to

$$\frac{x^2}{a^2}+\frac{4y^2}{b^2}\left(\frac{y^2}{b^2}-1\right)=0.$$

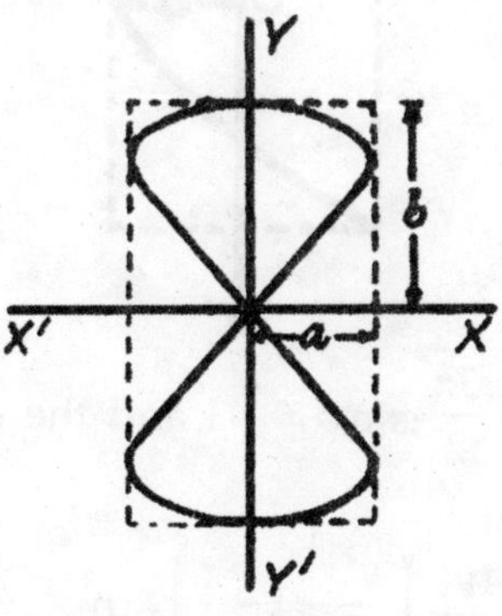

(b) when $\phi = \frac{\pi}{2}$, $\sin\phi = 1$ and the eq. (iii) reduces to

$$\left(\frac{x}{a}-1\right)^2 + \frac{4y^2}{b^2}\left(\frac{y^2}{b^2}-1+\frac{x}{a}\right) = 0$$

or $$\left(\frac{x}{a}-1\right)^2 + \frac{4y^4}{b^4} + \frac{4y^2}{b^2}\left(\frac{x}{a}-1\right) = 0$$

or $$\left[\left(\frac{x}{a}-1\right) + \frac{2y^2}{b^2}\right]^2 = 0.$$

This represents two coincident parabolas symmetrical about the x-axis each being $y^2 = -\frac{b^2}{2a}(x-a)$.

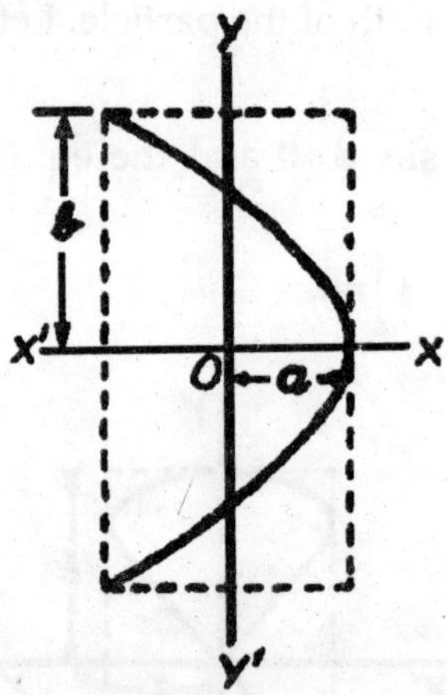

(c) When $\phi = \frac{3\pi}{2}$, $\sin\phi = -1$ and the eq. (iii) reduces to

$$\left(\frac{x}{a}+1\right)^2 + \frac{4y^2}{b^2}\left(\frac{y^2}{b^2}-1-\frac{x}{a}\right) = 0$$

or $$\left(\frac{x}{a}+1\right)^2+\frac{4y^4}{b^4}+\frac{4y^2}{b^2}\left(\frac{x}{a}+1\right)=0$$

or $$\left[\left(\frac{x}{a}+1\right)-\frac{2y^2}{b^2}\right]^2=0.$$

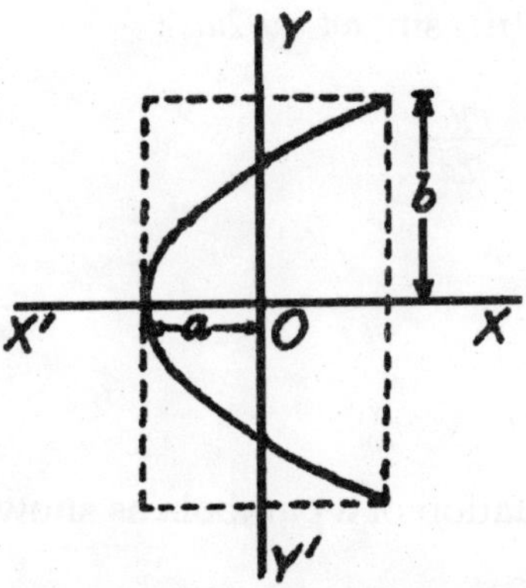

This again represents two coincident parabolas, each being $y^2 = \frac{b^2}{2a}$ (x+a) which is the appropriate Lissajous figure. If the frequencies depart slightly from the ratio 2 : 1, the form of the figures slowly varies.

Calculation of two Rectangular Simple Harmonic Motion

Let a particle be subjected to two simple harmonic motions of periods as well as amplitudes in ratio 1 : 2, and acting along the axes of x and y respectively. Let the x-motion lead over the y-motion by 90°, *i.e.* $\pi/2$ radian in phase. The equations of these motions would be

$$x = a \sin\left(2\,\omega t + \frac{\pi}{2}\right) \qquad ...(i)$$

and $$y=2a \sin \omega t, \qquad ...(ii)$$

where a and $2a$ are the amplitudes, and $2\pi/2\omega$ and $2\pi/\omega$ the periods respectively.

The equation of the resultant path of the particle is obtained by eliminating t between eq. (i) and (ii). From eq. (i), we have

$$\frac{x}{a} = \sin\left(2\,\omega t + \frac{\pi}{2}\right)$$

$$= \cos 2\,\omega t = 1 - 2\sin^2 \omega t$$

But from eq. (ii), $\sin\ \omega t = y/2a$.

$$\therefore \quad \frac{x}{a} = 1 - \frac{y^2}{2a^2}$$

or
$$\frac{y^2}{2a^2} = 1 - \frac{x}{a}$$

or
$$y^2 = 2a\ (a - x)$$

This is the equation of a parabola as shown in figure given below:

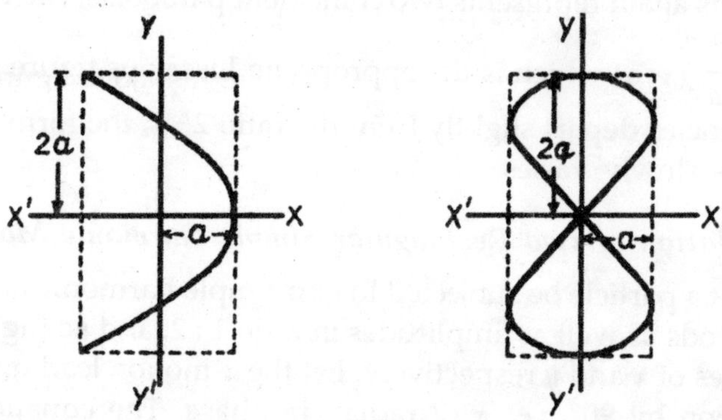

Let us now consider the case when they motion lead's over the x-motion by $\pi/2$. Now the equations of motions will be

$$x = a\ \sin^2\ \omega t \qquad ...(iii)$$

and
$$y = 2a\ \sin\left(\omega t + \frac{\pi}{2}\right). \qquad ...(iv)$$

From eq. (iii), we have

$$\frac{x}{a} = 2 \sin \omega t \cos \omega t$$

$$= 2\sqrt{(1 - \cos^2 \omega t)} \cos \omega t.$$

But from eq. (iv), $\frac{y}{2a} = \sin\left(\omega t + \frac{\pi}{2}\right) = \cos \omega t$.

Substituting for $\cos \omega t$ in the last eq. we get

$$\frac{x}{a} = 2\sqrt{\left(1 - \frac{y^2}{4a^2}\right)} \frac{y}{2a}$$

or $$\frac{x^2}{a^2} = \frac{y^2}{a^2}\left(1 - \frac{y^2}{4a^2}\right)$$

or $$x^2 + y^2\left(\frac{y^2}{4a^2} - 1\right) = 0.$$

Demonstration of Lissajous' Figures

The Lissajous' figures can be demonstrated by compounding two simple harmonic motions perpendicular to each other. This can be done by a number of methods:

Optical Method: Two electrically-maintained tuning forks F_1 and F_2 are taken with small mirrors M_1 and M_2 attached to their prongs. They are held in two planes at right angles to each other, as shown. A narrow and strong convergent beam of light taken from a source S and a lens L falls on the mirror M_1. It is reflected on to the mirror M_2 and finally produces a sharp spot of light O on a screen. If the forks are silent, the spot of light remains steady on the screen.

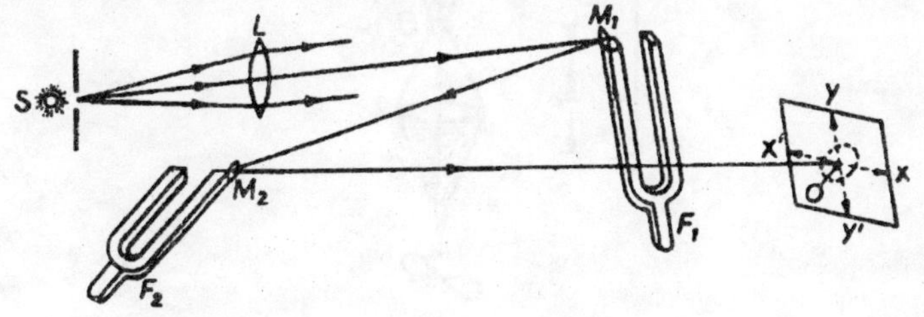

When only the fork F_1 is vibrated, the spot performs a S. H. M on the screen so rapidly that, due to persistence of vision, it appears to be drawn out into a vertical line along *YOY'* Similarly, when F_2 is vibrated, the spot is drawn out into a horizontal line along *XOX'*. When both forks are vibrated simultaneously, the motion of the spot of light follows the Lis 'ajous' figure obtained by compounding the two S. H. M. at right angles.

If the frequencies of the forks are exactly equal, the figure is constant in shape, its exact form depending on the relative phase of the component motions. If the frequencies are not exactly equal, the phase relation will pass slowly through all possible values and the figure will assume all forms until the form first observed is repeated.

Blackburn's Pendulum: A mechanical device to demonstrate the formation of Lissajous' figures is the Blackburn's pendulum. It consists of a string with its two ends attached to a rigid at points *M* and *N*. The string is cut at its centre from the ends so formed are attached to a heavy funnel *C*. The exit from the funnel is narrow so that when it is filled with sand, a fine stream falls on the ground. A clip *B* is placed so as to bring the two strings together. It can be slipped up and down on the strings.

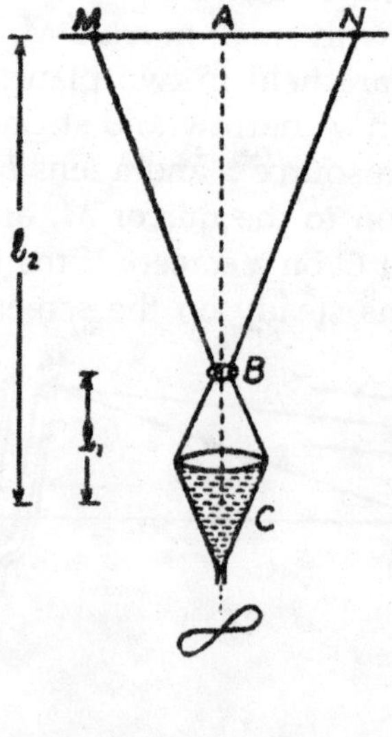

If the funnel is vibrated in the plane of the paper, the pendulum will have an effective length *BC*. If, however, it is vibrated perpendicular to the plane of paper, the pendulum will have an effective length *AC*. (The clip will have no effect on the period of vibration perpendicular to the plane of paper). If the funnel is pulled out slantwise and released, it moves under the action of two S.H.M.'s at right angles. Hence the stream of sand escaping from the funnel traces a Lissajous' figure on the ground.

Let BC=l_1 and $AC=l_2$. Then the periods of the two S. H. M.'s are given by

$$T_1 = 2\pi\sqrt{\left(\frac{l_1}{g}\right)}$$

and $$T_2 = 2\pi\sqrt{\left(\frac{l_2}{g}\right)},$$

$$\therefore \frac{T_1^2}{T_2^2} = \frac{l_1}{l_2}$$

By altering the position of the clip *B*, the lengths of the pendulums ll and l_2, may be varied to obtain different ratios of the periods.

For example, *if* $\frac{l_1}{l_2} = \frac{1}{4}$, we have

$$\frac{T_1}{T_2} = \frac{1}{2}$$

The sand, therefore, traces a curve having two loops. In a similar way, different types of figures may be obtained.

Vibration Microscope: This arrangement consists of a large standard tuning fork F_1 fixed horizontally on a support, and a small microscope *M* supported in a separate stand. The axis of the microscope is perpendicular to the plane of vibration of the fork. A small lens *L* attached to one of the prongs of the fork, serves as the objective lens of the microscope *M*. A second tuning fork F_2 is held vertically below F_1 so that its vibrations are perpendicular to those of F_1. The microscope is focused on a small point *D* marked on the top of a prong of F_2.

If the fork F_l alone is vibrated, the point *D* appears to be drawn out into a line parallel to the line *A* owing to the simple harmonic motion of the lens *L*. If the fork F_2 alone is vibrated, the dot appears as a line parallel to the line *B*. When both the forks are vibrated simultaneously, the point *D* appears to follow a Lissajous' figure appropriate to the ratio of frequencies of the forks. The frequency of F_2 can be altered either by filing its prongs at their ends (to increase the frequency) or by loading its prongs (to decrease the frequency) and different types of figures can be obtained.

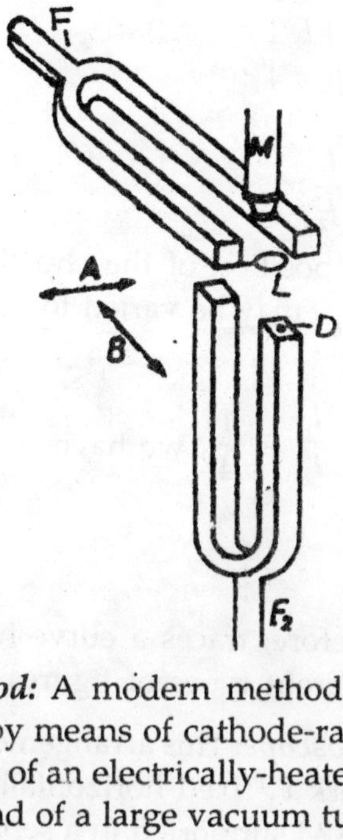

Electrical Method: A modern method for producing the Lissajous' figure is by means of cathode-ray oscilloscope. The oscilloscope consists of an electrically-heated cathode filament C at the narrower end of a large vacuum tube, whose broader end *F* is painted by a fluorescent material on the inside. The filament *C* emits a beam of electrons. It is surrounded by a shield *S* which is maintained at a small negative potential with respect to the filament so as to concentrate the electron-beam along its axis.

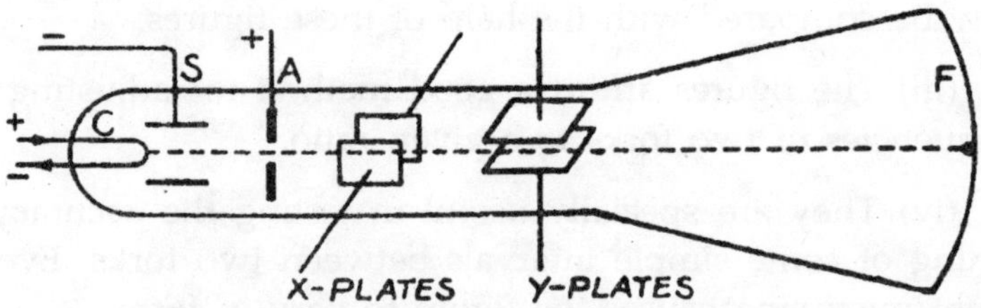

The beam is accelerated by a positive plate *A* (anode) and passes through a hole made in it. It then passes through two pairs of parallel plates placed vertically and horizontally and known as X-plates and Y-plates respectively. Finally, it falls on the fluorescent screen *F and* produces a luminous spot on it.

If an alternating potential difference is applied across either the X- or the Y-plates, then the electron-beam will be acted upon by a simple harmonic force. Hence the movement of the spot on the screen will be simple harmonic and will be shown as a straight line on the screen. When alternating potentials are applied simultaneously to the *X*- and Y-plates, the electron-beam moves under two S. H. M.'s at right angles to each other and the spot of light traces out a Lissajous' figure on the screen. If the two potentials are of the same frequency, then by varying their amplitudes and phases we can obtain a straight line, a circle or an ellipse on the screen. If the potentials have different, frequencies, then very interesting patterns may result.

This is the best method for producing Lissajous' figures since the very small inertia of the electron-beam allows it to respond faithfully to the two perpendicular simple harmonic motions.

Uses of Lissajous' Figures

The Lissajous' figures have a number of practical uses:

(i) If the frequencies of two vibrating systems are commensurable, *i.e.*, in a whole number ratio as 1 : 1, 2 : 1. 3 : 1, the ratio may be read off at once from the steady figure they produce.

(ii) The frequencies of two systems, if nearly commensurate, may be compared with the help of these figures.

(iii) The figures afford a good method for adjusting the frequencies of two forks to a given ratio.

(iv) They are specially useful in testing the accuracy of tuning of some simple intervals between two forks. Even a slight mistiming causes the figure to vary in form.

(v) The principle of Lissajous' figures may be employed to investigate how the period of a rod fixed at one end, varies with the length of the rod.

(vi) Helmholtz used the principle to investigate the vibration of a violin string.

Determination of Frequency of Tuning Fork

The Lissajous' figures provide an important method of comparing the frequencies of two tuning forks.

The tuning forks are arranged to produce Lissajous' figures which are carefully studied. There are three important cases:

Case (i)—When the Frequencies are in a Whole Number Ratio: Let the frequencies of the two forks be n_1 and n_2; such that n_1/n_2=l, 2, 3, etc. In this case a steady figure is obtained. Suppose it is of the form shown in figure below. Let us draw the coordinate axes across the figure.

Suppose the fork of frequency n_1 is vibrating along the X-axis and the fork of frequency n_2 along the Y-axis. Suppose the complete figure is traced in *t* second. Clearly, during this time the number of vibrations along the X-axis will be n_1t, while the number along the Y-axis will be n_2t.

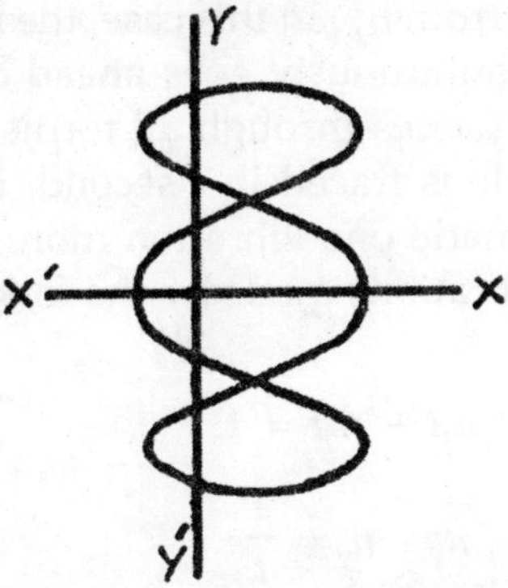

Now, a vibrating point passes through any point in its path twice in one complete vibration. Hence during n_1t vibrations along the X-axis, the figure will cut the Y-axis $2n_1t$ times. Similarly, it will cut the X-axis $2n_2t$ times. Thus

$$\frac{\text{number of times the figure cuts the } Y\text{ - axis}}{\text{number of times the figure cuts the X - axis}}$$

$$= \frac{2n_1t}{2n_2t} = \frac{n_1}{n_2} = \frac{\text{frequency of X - vibrations}}{\text{frequency of } Y\text{ - vibrations}}$$

Hence if p_y and p_x be the number of times the figure cuts the *Y*- and X-axis respectively, we have

$$\frac{n_1}{n_2} = \frac{p_y}{p_x}.$$

Thus, if the number p_x and p_y are counted, the ratio of the frequencies of the tuning forks is obtained. In figure above we have

$$py = 6,\ p2 = 2.$$

$$\therefore \quad \frac{n_1}{n_2} = \frac{6}{2} = \frac{3}{1}$$

Therefore the ratio of the frequencies is 3 : 1.

Case (ii)—Frequencies Nearly Equal: Suppose the frequency of one fork is n_1 and that of the other n_2, such that n_1 is only slightly different from n_2. In this case, the motion of the higher frequency fork continuously goes ahead of the other and the Lissajous' figure passes through all forms of a cycle. Suppose the complete cycle is traced in t second. During t second the higher fork has made one vibration more than the lower. But the number of vibrations made by the forks in t second are n_1t and n_2t. Thus

$$n_1t \sim n_2t = 1$$

or
$$n_1 \sim n_2 = \frac{1}{t}$$

or
$$n_1 = n_2 \pm \frac{1}{t}$$

Thus if n_2=320 and t = 5 second then n_1=320.2 or 319.8. To decide which of these two values n_1 have, a piece of wax is attached to the prong of the fork n_1 which lowers its frequency slightly. The experiment is repeated and the new time of cycle is noted. If it is more than 5 second then n=320.2; but if is less than 5 second then n_1=319.8.

[If t has increased, it means that on lowering n_1, the difference $n_1 \sim n_2$ has decreased. Hence n_1 must have been higher than n_2. But if t has decreased, it means that on lowering n_1, the difference $n_1 \sim n_2$ has increased. Hence n_1 must have been lower than n_2].

Case (Hi)—Frequency nearly 2 : 1: When the frequencies of the two forks are in the ratio nearly 2 : 1, the Lissajous' figure changes from parabola to the figure of eight and back to same parabola during the phase change of 2π. Let t second be the time during which a complete cycle of Lissajous' figure is traced. Obviously, in this time, the higher fork (n_1) has made one vibration more or one vibration less than twice the number made by the lower fork (n_2). Hence

$$n_1t \sim 2n_2t = 1$$

or
$$n_1 \sim 2n_2 = \frac{1}{t}$$

or $$n_1 = 2n_2 \pm \frac{1}{t}$$

Knowing n_2 and t, n_1 can be determined. To decide whether $n_1 > 2n_2$ or $n_1 < 2n_2$, the higher fork n_1 is slightly loaded with wax and the time for the complete cycle of figures is again observed. If the time has increased then $n_1 > 2n_2$, but if time has decreased then $n_1 < 2n_2$.

Problems

1. Two toning forks produce Lissajous' figures of the shape of straight line, ellipse and finally to the same straight line in 2 seconds. If one of the toning forks has a frequency of 190 vib/sec, find the possible frequencies of the other.

Solution: The Lissajous' figures are straight line and ellipse. Hence the frequencies of the two forks must nearly be in the ratio 1:1.

As the cycle of figures is traced in 2 seconds, the difference in frequencies must be $\frac{1}{2}$ sec^{-1}. Therefore, the other fork must have one of the following frequencies:

$$100 \pm \frac{1}{2} = 100.5, 99.5.$$

2. The Lissajous' figure in the case of two toning forks is a parabola. Prove that the frequencies are in the ratio 1 : 2.

Solution: Let us draw coordinate axes across the given parabola. Let one fork of frequency n_1 be vibrating along the X-axis and the other of frequency n_2 be vibrating along the *Y*-axis. Then if the parabola cuts the *Y*-axis p_y times and the X-axis p_x times, we have

$$\frac{n_1}{n_2} = \frac{p_y}{p_x}$$

Here p_y=2 and p_x=1. Therefore

$$\frac{n_1}{n_2} = \frac{2}{1}$$

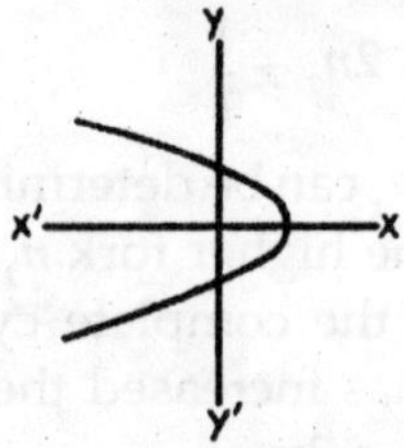

3. Two tuning forks of approximately equal frequencies produce Lissajous' figures that go through a cycle of changes in 15 seconds. When one fork is loaded with wax, the cycle of changes of Lissajous' figures takes 10 seconds. If the second tuning fork has a frequency of 400, find the frequency of the first fork before and after loading.

Solution: Let n_1 and n_2 be the frequencies of the two forks. Before loading, the cycle of Lissajous' figures takes 15 seconds. This shows that in 15 seconds one fork makes one complete vibration more (or less) than the other. Thus

$$n_1 \times 15 \sim n_2 \times 15 = 1$$

or
$$n_1 \sim n_2 = 1/15$$

or
$$n_1 = n_2 \pm 1/15$$

But $n_2 = 400$.

$$\therefore n_1 = 400 \pm 1/15$$
$$= 400.067 \text{ or } 399933.$$

On loading the first fork with wax, the time of cycle decreases to 10 second. Hence the frequency of the first fork after loading

$$n_1 = n_2 \pm 1/10$$
$$= 400 \pm 1/10$$
$$= 4001 \text{ or } 399.9.$$

But, on loading, the frequency will decrease. Hence its frequency before loading is 399.933 and after loading 399.9.

4. Two tuning forks A and B of nearly equal frequencies are employed to produce Lissajous' figures. On slightly loading A, the cycle of change of figures slows down from 15 sec to 20 sec. If the frequency of B is 256, find the frequency of A before and after loading.

$$\left[\text{Ans. } 256+\frac{1}{15}, 236+\frac{1}{20}\right]$$

5. Two tuning forks whose frequencies are approximately in the ratio 2 : 1 are employed to produce Lissajous' figures and it is observed that the figures go through a cycle of changes in 15 seconds. On loading slightly the fork of higher pitch, the figures go through a cycle of changes in 10 seconds. If the fork of the lower pitch has a frequency of 256, find the frequency of the other fork before and after loading.

Solution: Let n_1 and n_2, be the frequencies of the two forks, n_1 being the higher. Before loading, the figures go through a cycle of changes in 15 seconds. This shows that in 15 seconds the higher fork makes one complete vibration more (or less) than twice the number of vibrations made by the lower. Thus

$$n_1 \times 15 \sim 2\,(n_2 \times 15) = 1$$

or $$n_1 \sim 2n_2 = 1/15$$

or $$n_1 = 2n_2 \pm 1/15.$$

But $n_2 = 256$.

$$\therefore\ n_1 = 2 \times 256 \pm 1/15$$

$$= 512 \pm 1/15$$

$$= 512.067 \text{ or } 511.933.$$

When the fork of higher pitch is slightly loaded, the time of the cycle decreases to 10 seconds. Hence the frequency of the higher fork after loading

$$n_1' = 2n_2 \pm 1/10$$

$$= 2 \times 256 \pm 1/10$$

$$= 512 \pm 1/10$$

$$= 512.1 \text{ or } 511.9.$$

But, on loading the fork the frequency will decrease. Hence its frequency before loading is 511.933 and after loading 511.9.

6. *Two tuning forks produce Lissajoas' figures. The figure changes in form from a parabola to a figure of eight and again to a parabola, the whole cycle occupying 6 seconds. If the frequency of one fork is 100, find the possible frequencies of the other.*

Solution: Since the figure changes in form from a parabola to a figure of eight and again to a parabola, the frequency ratio must be 'nearly' 2 : 1. The frequency of one fork is 100. Hence the frequency of the other must be 'nearly' 50 or 'nearly' 200.

Case I—When the unknown frequency n is lower than the known frequency 100: Since the cycle of figures takes 6 seconds, the unknown fork of lower frequency n will make $6n$ vibrations in this time, while the known fork will make $2\,(6n)\pm 1$ vibrations during the same time. But, this should be equal to 6×100. because the frequency of the known fork is 100.

$$\therefore \quad 2(6n) \pm 1 = 6 \times 100$$

or
$$12n \pm 1 = 600$$

or
$$12n = 600 \mp 1$$

or
$$n = 50 \mp 1/12$$

Case II—When the unknown frequency n is higher than the known frequency 100: The known fork of lower frequency 100 will make 600 vibrations during the interval of 6 seconds, after which the cycle of figures is repeated. Clearly, in this case the unknown fork will make $2(600)\pm 1$ vibrations in the same time. This should be equal to $6n$, where n is the unknown frequency. Hence we have

$$6n = 2\,(600) \pm 1$$

$$\therefore \quad n = 200 \pm \tfrac{1}{6}$$

The possible frequencies of the unknown fork are

$$(50 \pm 1/12) \text{ and } (200 \pm 1/6).$$

4

Harmonic Oscillator

In classical mechanics, a harmonic oscillator is a system which, when displaced from its equilibrium position, experiences a restoring force F proportional to the displacement x according to Hooke's law:

$$F = -kx$$

where k is a positive constant.

If F is the only force acting on the system, the system is called a simple harmonic oscillator, and it undergoes simple harmonic motion: sinusoidal oscillations about the equilibrium point, with a constant amplitude and a constant frequency (which does not depend on the amplitude).

If a frictional force (damping) proportional to the velocity is also present, the harmonic oscillator is described as a damped oscillator. In such situation, the frequency of the oscillations is smaller than in the non-damped case, and the amplitude of the oscillations decreases with time.

If an external time-dependent force is present, the harmonic oscillator is described as a driven oscillator.

Mechanical examples include pendula (with small angles of displacement), masses connected to springs, and acoustical systems. Other analogous systems include electrical harmonic oscillators.

The simple harmonic oscillator has no driving force, and no friction (damping), so the net force is just:

$F = -kx$

Using Newton's Second Law

$F = ma = -kx$

The acceleration, a is equal to the second derivative of x.

$$m\frac{d^2x}{dt^2} = -kx$$

If we define, then the equation can be written as follows,

$$\frac{d^2x}{dt^2} + \omega_0{}^2 x = 0$$

We observe that:

$$\frac{d^2x}{dt^2} = \ddot{x} = \frac{d\dot{x}}{dt}\frac{dx}{dx} = \frac{d\dot{x}}{dx}\frac{dx}{dt} = \frac{d\dot{x}}{dt}\dot{x}$$

and substituting

$$\frac{d\dot{x}}{dt}\dot{x} + \omega_0{}^2 x = 0 \quad d\dot{x}.\,\dot{x} + \omega_0{}^2 x\,.\,dx = 0$$

integrating

$$\dot{x}^2 + \omega_0{}^2 x^2 = K$$

where K is the integration constant, set $K = (A\ \omega_0)^2$

$$\dot{x}^2 = A^2\omega_0{}^2 - \omega_0{}^2 x^2 \qquad \dot{x} = \pm\,\omega_0\sqrt{A^2 - x^2}$$

$$\frac{dx}{\pm\sqrt{A^2 - x^2}} = \omega_0 dt$$

integrating, the results (including integration constant φ) are

$$\begin{cases} \arcsin\dfrac{x}{A} = \omega_0 t + \phi \\ \arccos\dfrac{x}{A} = \omega_0 t + \phi \end{cases}$$

and has the general solution

$$x = A\cos(\omega_0 t + \phi)$$

where the amplitude A and the phase ϕ are determined by the initial conditions.

Alternatively, the general solution can be written as

$$x = A\sin(\omega_0 t + \phi)$$

where the value of ϕ is shifted by $\frac{\pi}{2}$ relative to the previous form;

or as

$$x = A\sin\omega_0 t + B\cos\omega_0 t$$

where A and B are the constants which are determined by the initial conditions, instead of A and ϕ in the previous forms.

The frequency of the oscillations is given by

$$f = \frac{\omega_0}{2\pi}$$

The kinetic energy is

$$T = \frac{1}{2}m\left(\frac{dx}{dt}\right)^2 = \frac{1}{2}kA^2\sin^2(\omega_0 t + \phi)$$

and the potential energy is

$$U = \frac{1}{2}kx^2 = \frac{1}{2}kA^2\cos^2(\omega_0 t + \phi)$$

so the total energy of the system has the constant value

$$E = \frac{1}{2}kA^2$$

Driven Harmonic Oscillator

A driven harmonic oscillator satisfies the non-homogeneous second order linear differential equation

$$\frac{d^2x}{dt^2} + \omega_0{}^2 x = A_0\cos(\omega t)$$

where *AO* is the driving amplitude and ω is the driving frequency for a sinusoidal driving mechanism. This type of system appears in AC LC (inductor-capacitor) circuits and idealised spring systems lacking internal mechanical resistance or external air resistance.

Damped Harmonic Oscillator

A damped harmonic oscillator satisfies the second order differential equation

$$\frac{d^2x}{dt^2} + \frac{b}{m}\frac{dx}{dt} + \omega_0{}^2 x = 0,$$

where *b* is an experimentally determined damping constant satisfying the relationship $F = -bv$. An example of a system obeying this equation would be a weighted spring underwater if the damping force exerted by the water is assumed to be linearly proportional to *v*.

The frequency of the damped harmonic oscillator is given by

$$\omega_1 = \sqrt{w_0^2 = R_m^2}$$

where

$$R_m = \frac{b}{2m}$$

Damped, Driven Harmonic Oscillator

This satisfies the equation

$$m\frac{d^2x}{dt^2} + r\frac{dx}{dt} + kx = F_0 \cos(\omega t)$$

The general solution is a sum of a transient (the solution for damped undriven harmonic oscillator, homogeneous ODE) that depends on initial conditions, and a steady state (particular solution of the non-homogenous ODE) that is independent of initial conditions and depends only on driving frequency, driving force, restoring force, damping force,

The steady-state solution is

$$x(t) = \frac{F_0}{Z_m \omega} \sin(\omega t - \phi)$$

where

$$Z_m = \sqrt{r^2 + \left(\omega m = \frac{k}{\omega}\right)^2}$$

is the absolute value of the impedance or linear response function

$$Z = r + i\left(\omega m = \frac{k}{\omega}\right)$$

and

$$\phi = \arctan\left(\frac{\omega m = \frac{k}{\omega}}{r}\right)$$

is the phase of the oscillation relative to the driving force.

One might see that for a certain driving frequency, ω, the amplitude (relative to a given *F*0) is maximal. This occurs for the frequency

$$\omega_r = \sqrt{\frac{k}{m} - 2\left(\frac{r}{2m}\right)^2}$$

and is called resonance of displacement.

In summary: at a steady state the frequency of the oscillation is the same as that of the driving force, but the oscillation is phase-offset and scaled by amounts that depend on the frequency of the driving force in relation to the preferred (resonant) frequency of the oscillating system.

Example: RLC circuit.

Full mathematical definition

Most harmonic oscillators, at least approximately, solve the differential equation:

$$\frac{d^2x}{dt^2} + \frac{b}{m}\frac{dx}{dt} + \omega_0^{\,2}x = A_0\cos(\omega t)$$

where t is time, b is the damping constant, ω_0 is the characteristic angular frequency, and $A_0\cos(\omega t)$ represents something driving the system with amplitude A_0 and angular frequency ω. x is the measurement that is oscillating; it can be position, current, or nearly anything else. The angular frequency is related to the frequency, *f*, by

$$f = \frac{\omega}{2\pi}$$

Important Terms

- *Amplitude:* Maximal displacement from the equilibrium.
- *Period:* The time it takes the system to complete an oscillation cycle. Inverse of frequency.
- *Frequency:* The number of cycles the system performs per unit time (usually measured in hertz = 1/s).
- *Angular Frequency:* $\omega = 2\pi f$
- *Phase:* How much of a cycle the system completed (system that begins is in phase zero, system which completed half a cycle is in phase π).
- *Initial Conditions:* The state of the system at $t = 0$, the beginning of oscillations.

Simple Harmonic Oscillator

A simple harmonic oscillator is simply an oscillator that is neither damped nor driven. So the equation to describe one is:

$$\frac{d^2x}{dt^2} + \omega_0{}^2 x = 0.$$

Physically, the above never actually exists, since there will always be friction or some other resistance, but two approximate examples are a mass on a spring and an LC circuit.

In the case of a mass attached to a spring, Newton's Laws, combined with Hooke's law for the behaviour of a spring, states that:

$-kx = ma$

where k is the spring constant

- m is the mass

- x is the position of the mass
- a is its acceleration.

Because acceleration a is the second derivative of position x, we can rewrite the equation as follows:

$$-kx = m\frac{d^2x}{dt^2}$$

The most simple solution to the above differential equation is

$$x = A\cos(\omega t + \delta)$$

and the second derivative of that is

$$\frac{d^2x}{dt^2} = -A\omega^2\cos(\omega t + \delta)$$

where A is the amplitude, δ is the phase shift, and ω is the angular frequency.

Plugging these back into the original differential equation, we have:

$$-Ak\cos(\omega t + \delta) = -Am\omega^2\cos(\omega t + \delta)$$

Then, after dividing both sides $-A\cos(\omega t + \delta)$ by we get:

$$k = m\omega^2$$

or, as it is more commonly written:

$$\omega = \sqrt{\frac{k}{m}}$$

The above formula reveals that the angular frequency ω of the solution is only dependent upon the physical characteristics of the system, and not the initial conditions (those are represented by A and δ). We will label this ω as ω_0 from now on. This will become important later.

Universal Oscillator Equation

The equation

$$\frac{d^2q}{d\tau^2}+2\zeta\frac{dq}{d\tau}+q=0$$

is known as the universal oscillator equation since all second order linear oscillatory systems can be reduced to this form. This is done through non-dimensionalisation.

If the forcing function is $f(t) = \cos(\omega t) = \cos(\omega t_c \tau) = \cos(\omega \tau)$, where $\omega = \omega t_c$, the equation becomes

$$\frac{d^2q}{d\tau^2}+2\zeta\frac{dq}{d\tau}+q=\cos(\omega\tau)$$

The solution to this differential equation contains two parts, the "transient" and the "steady state".

Transient Solution

The solution based on solving the ordinary differential equation is for arbitrary constants c_1 and c_2 is

$$q_t(\tau)=\begin{cases} e^{-\zeta\tau}\left(c_1e^{\tau}\sqrt{\zeta^2-1}+c_2e^{-\tau}\sqrt{\zeta^2-1}\right) & \zeta>1(\textit{overdamping}) \\ e^{-\zeta\tau}(c_1+c_2\tau)=e^{-\tau}(c_1+c_2\tau) & \zeta=1(\textit{critical damping}) \\ e^{-\zeta\tau}\left[c_1\cos\left(\sqrt{1-\zeta^2\tau}\right)+c_2\sin\left(\sqrt{1-\zeta^2\tau}\right)\right] & \zeta<1(\textit{underdammping}) \end{cases}$$

The transient solution is independent of the forcing function. If the system is critically damped, the response is independent of the damping.

Steady-state Solution

Apply the "complex variables method" by solving the auxiliary equation below and then finding the real part of its solution:

$$\frac{d^2q}{d\tau^2}+2\zeta\frac{dq}{d\tau}+q=\cos(\omega\tau)+i\sin(\omega\tau)=e^{i\omega\tau}$$

Supposing the solution is of the form

$$q_s(\tau) = Ae^{i(\omega\tau+\phi)}$$

Its derivatives from zero to 2nd order are

$$q_s = Ae^{i(\omega\tau+\phi)}, \frac{dq_s}{d\tau} = i\omega Ae^{i(\omega\tau+\phi)}, \frac{d^2q_s}{d\tau^2} = -\omega^2 Ae^{i(\omega\tau+\phi)}$$

Substituting these quantities into the differential equation gives

$$-\omega^2 Ae^{i(\omega\tau+\phi)} + 2\zeta i\omega Ae^{i(\omega\tau+\phi)} + Ae^{i(\omega\tau+\phi)} = (-\omega^2 A + 2\zeta iwA + A)e^{i(\omega\tau+\phi)} = e^{i\omega\tau}$$

Dividing by the exponential term on the left results in

$$-\omega^2 A + 2\zeta i\omega A + A = e^{-e\phi} = \cos\phi - i\sin\phi$$

Equating the real and imaginary parts results in two independent equations

$$A(1-\omega^2) = \cos\phi \qquad 2\zeta\omega A = -\sin\phi.$$

Amplitude part

Squaring both equations and adding them together gives

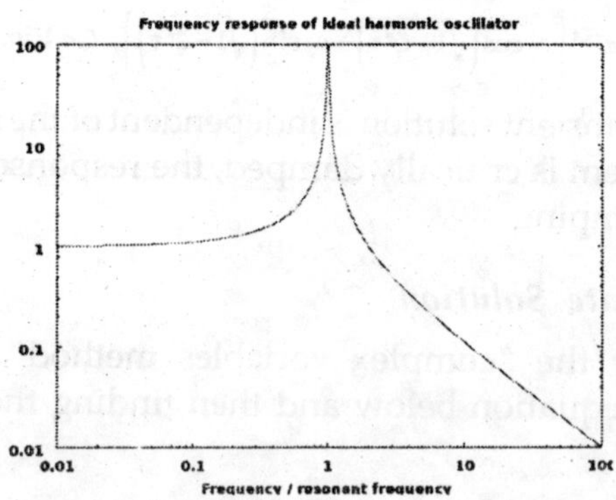

$$\left.\begin{aligned} A^2(1-\omega^2)^2 &= \cos^2\phi \\ (2\zeta\omega A)^2 &= \sin^2\phi \end{aligned}\right\} \Rightarrow A^2\left[\left(1-\omega^2\right)^2 + (2\zeta\omega)^2\right] = 1$$

By convention the positive root is taken since amplitude is usually considered a positive quantity. Therefore,

$$A = A(\zeta,\omega) = \frac{1}{\sqrt{(1-\omega^2)^2 + (2\zeta\omega)^2}}$$

This amplitude function is particularly important in the analysis and understanding of the frequency response of second-order systems. Note that the variables in these equations ought to be identified before showing the equation.

Phase part

To solve for φ, divide both equations to get

$$\tan\phi = -\frac{2\zeta\omega}{1-\omega^2} = \frac{2\zeta\omega}{\omega^2-1} \Rightarrow \phi \equiv \phi(\zeta,\omega) = \arctan\left(\frac{2\zeta\omega}{\omega^2-1}\right)$$

This phase function is particularly important in the analysis and understanding of the frequency response of second-order systems.

Full Solution

Combining the amplitude and phase portions results in the steady-state solution

$$q_s(\tau) = A(\zeta,\omega)\cos\big(\omega\tau + \phi(\zeta,\omega)\big) = A\cos(\omega\tau + \phi)$$

The solution of original universal oscillator equation is a superposition (sum) of the transient and steady-state solutions

$$q(\tau) = q_t(\tau) + q_s(\tau)$$

Relationship to RLC circuit

Comparing a mechanical harmonic oscillator with a series RLC circuit or parallel RLC circuit, the following correspond:

- F (force) $\Leftrightarrow$ V (electric potential) $\Leftrightarrow$ I (electric current)
- x (position) $\Leftrightarrow$ Q (charge) $\Leftrightarrow$ Φ (magnetic flux)

- k (spring constant) $\Leftrightarrow$ (electrical elastance (reciprocal of capacitance) $\Leftrightarrow \frac{1}{L}$ (reciprocal of inductance)
- v (velocity) $\Leftrightarrow I$ (electric current) $\Leftrightarrow V$ (electric potential)
- b (damping factor) $\Leftrightarrow R$ (electrical resistance) $\Leftrightarrow \frac{1}{R}$ (reciprocal of electrical resistance)
- a (acceleration) $\Leftrightarrow \frac{dI}{dt}$ (rate of change of current) $\Leftrightarrow \frac{dV}{dt}$ (rate of change of electric potential)
- m (mass) $\Leftrightarrow L$ (inductance) $\Leftrightarrow C$ (capacitance)

Applications

The problem of the simple harmonic oscillator occurs frequently in physics because of the form of its potential energy function:

$$V(x) = \frac{1}{2}kx^2$$

Given an arbitrary potential energy function $V(x)$, one can do a Taylor expansion in terms of x around an energy minimum ($x = x_0$) to model the behaviour of small perturbations from equilibrium.

$$V(x) = V(x_0) + (x - x_0)V'(x_0) + \frac{1}{2}(x - x_0)^2 V^{(2)}(x_0) + O(x - x_0)^3$$

Because $V(x_0)$ is a minimum, the first derivative evaluated at x_0 must be zero, so the linear term drops out:

$$V(x) = V(x_0) + \frac{1}{2}(x - x_0)^2 V^{(2)}(x_0) + O(x - x_0)^3$$

The constant term is arbitrary and thus may be dropped, and a coordinate transformation allows the form of the simple harmonic oscillator to be retrieved:

$$V(x) \approx \frac{1}{2}x^2V^{(2)}(0) = \frac{1}{2}kx^2$$

Thus, given an arbitrary potential energy function $V(x)$ with a non-vanishing second derivative, one can use the solution to the simple harmonic oscillator to provide an approximate solution for small perturbations around the equilibrium point.

Examples

Simple Pendulum: Assuming no damping and small amplitudes, the differential equation governing a simple pendulum is

$$\frac{d^2\theta}{dt^2} + \frac{g}{l}\theta = 0$$

The solution to this equation is given by:

$$\theta(t) = \theta_0 \cos\left(\sqrt{\frac{g}{l}}t\right) \qquad |\theta_0| \ll 1$$

where θ_0 is the largest angle attained by the pendulum. The period, the time for one complete oscillation , is given by 2π divided by whatever is multiplying

$$T_0 = 2\pi\sqrt{\frac{l}{g}} \qquad |\theta_0| \ll 1$$

Swinging Pendulum

Simple harmonic motion can in some cases be considered to be the one-dimensional projection of two-dimensional circular motion. Consider a long pendulum swinging over the turntable of a record player. On the edge of the turntable there is an object. If the object is viewed from the same level as the turntable, a projection of the motion of the object seems to be moving backwards and forwards on a straight line.

It is possible to change the frequency of rotation of the turntable in order to have a perfect synchronisation with the motion of the pendulum. The angular speed of the turntable is the pulsation of the pendulum.

In general, the pulsation-also known as angular frequency, of a straight-line simple harmonic motion is the angular speed of the corresponding circular motion.

Therefore, a motion with period T and frequency f=1/T has pulsation

$$\omega = 2\pi f = \frac{2\pi}{T}$$

In general, pulsation and angular speed are not synonymous. For instance the pulsation of a pendulum is not the angular speed of the pendulum itself, but it is the angular speed of the corresponding circular motion.

Spring-mass System

When a spring is stretched or compressed by a mass, the spring develops a restoring force. Hooke's law gives the relationship of the force exerted by the spring when the spring is compressed or stretched a certain length:

$$F(t) = kx(t)$$

where F is the force, k is the spring constant, and x is the displacement of the mass with respect to the equilibrium position.

This relationship shows that the distance of the spring is always opposite to the force of the spring.

By using either force balance or an energy method, it can be readily shown that the motion of this system is given by the following differential equation:

$$m\frac{d^2}{dt^2}x(t) + kx(t) = 0$$

If the initial displacement is A, and there is no initial velocity, the solution of this equation is given by:

$$x(t) = A\cos\left(\left(\sqrt{k/m}\right)t\right)$$

Energy Variation in the Spring-damper System

In terms of energy, all systems have two types of energy, potential energy and kinetic energy. When a spring is stretched or compressed, it stores elastic potential energy, which then is transferred into kinetic energy. The potential energy within a spring is determined by the equation $U = 1/2kx^2$.

When the spring is stretched or compressed, kinetic energy of the mass gets converted into potential energy of the spring. By conservation of energy, assuming the datum is defined at the equilibrium position, when the spring reaches its maximum potential energy, the kinetic energy of the mass is zero. When the spring is released, the spring will try to reach back to equilibrium, and all its potential energy is converted into kinetic energy of the mass.

Q Factor

The Q factor or quality factor compares the time constant for decay of an oscillating physical system's amplitude to its oscillation period. Equivalently, it compares the frequency at which a system oscillates to the rate at which it dissipates its energy. A higher *Q* indicates a lower rate of energy dissipation relative to the oscillation frequency. For example, a pendulum suspended from a high-quality bearing, oscillating in air, would have a high *Q*, while a pendulum immersed in oil would have a low one.

Generally *Q* is defined to be

$$Q = \omega \frac{Energy\,Stored}{Power\,Loss}$$

where ω is defined to be the angular frequency of the circuit (system), and the energy stored and power loss are properties of a system under consideration.

Usefulness of 'Q'

The Q factor is particularly useful in determining the qualitative behaviour of a system. For example, a system with Q less than or equal to 1/2 cannot be described as oscillating at all, instead the system is said to be in an overdamped ($Q < 1/2$) or critically damped ($Q = 1/2$) state. However, if $Q > 1/2$, the system's amplitude oscillates, while simultaneously decaying exponentially. This regime is referred to as underdamped.

Physical Interpretation of Q

Physically speaking, Q is 2ð times the ratio of the total energy stored divided by the energy lost in a single cycle.

Equivalently (for large values of Q), the Q factor is approximately the number of oscillations required for a freely oscillating system's energy to fall off to $1/e^{2\pi}$, or about 1/535, of its original energy.

When the system is driven by a sinusoidal drive, its resonant behaviour depends strongly on Q. Resonant systems respond to frequencies close to their natural frequency much more strongly than they respond to other frequencies. A system with a high Q resonates with a greater amplitude (at the resonant frequency) than one with a low Q factor, and its response falls off more rapidly as the frequency moves away from resonance. Thus, a radio receiver with a high Q would be more difficult to tune with the necessary precision, but would do a better job of filtering out signals from other stations that lay nearby on the spectrum. The width of the resonance is given by

$$\Delta f = \frac{f_0}{Q},$$

where f_0 is the resonant frequency, and Δf, the bandwidth, is the width of the range of frequencies for which the energy is at least half its peak value.

Electrical Systems

For an electrically resonant system, the Q factor represents the effect of electrical resistance and, for electromechanical resonators such as quartz crystals, mechanical friction.

In a series RLC circuit, and in a tuned radio frequency receiver (TRF) the Q factor is:

$$Q = \frac{1}{R}\sqrt{\frac{L}{C}},$$

where R, L and C are the resistance, inductance and capacitance of the tuned circuit, respectively.

In a parallel RLC circuit, Q is equal to the reciprocal of the above expression.

Often for an electrical system the response can most easily be measured as an amplitude (voltage or a current), rather than energy or power. Since power and energy are proportional to the square of the amplitude of the oscillation, the bandwidth on an amplitude-frequency graph should be measured to $\frac{1}{\sqrt{2}}$ of the peak (approximately –3 dB), rather than 1/2 (–6 dB). A point where this occurs is known as a corner frequency, a cutoff frequency, a half-power point, or a break point.

Mechanical Systems

For a single damped mass-spring system, the Q factor represents the effect of mechanical resistance.

$$Q = \frac{\sqrt{MK}}{R}$$

where M is the mass, K is the spring constant, and R is the mechanical resistance, defined by the equation $F_{damping} = -Rv$, where v is the velocity.

Optical Systems

In optics, the Q factor of a resonant cavity is given by

$$Q = \frac{2\pi f_o \varepsilon}{P},$$

where f_o is the resonant frequency, ε is the stored energy in the cavity, and $P = -\frac{dE}{dt}$ is the power dissipated. The optical Q is equal to the ratio of the resonant frequency to the bandwidth of the cavity resonance. The average lifetime of a resonant photon in the cavity is proportional to the cavity's Q. If the Q factor of a laser's cavity is abruptly changed from a low value to a high one, the laser will emit a pulse of light that is much more intense than the laser's normal continuous output. This technique is known as Q-switching.

Normal Mode

A normal mode in an oscillating system is the frequency at which a deformable structure will oscillate when disturbed. Normal modes are also known as natural frequencies or resonant frequencies. There is a set of these frequencies that are unique to each structure.

It is common to use a spring-mass system to illustrate a deformable structure. When such a system is excited at one of these natural frequencies, all of the masses move at the same frequency. The phases of the masses are either exactly the same or exactly opposite. The practical significance of this can be illustrated by a mass-spring model of a building. If an earthquake excites the system near one of the natural frequencies, the displacement of one floor with respect to another will be maximum. Obviously, buildings can only withstand this displacement up to a certain point. Being able to model a building and find its normal modes is an easy way to check the safety of a building's design. The concept of normal modes also finds application in wave theory, optics and quantum mechanics.

Example-normal Modes of Coupled Oscillators

Consider two bodies (not affected by gravity), each of mass M, attached to three springs with stiffness K. They are attached in the following manner:

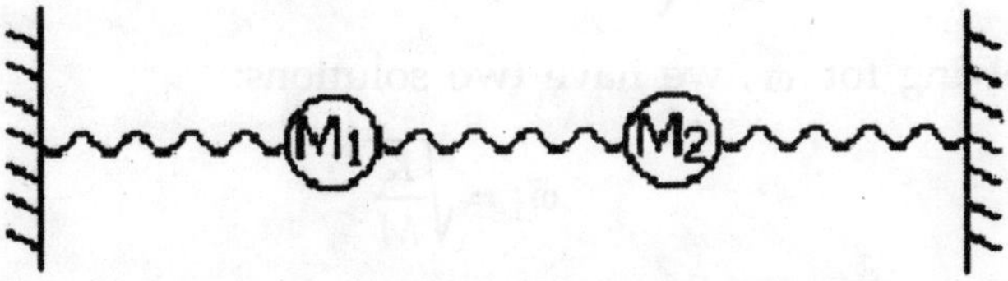

where the edge points are fixed and cannot move. We'll use $x_1(t)$ to denote the displacement of the leftmost mass, and $x_2(t)$ to denote the displacement of the rightmost.

If we denote the second derivative of $x(t)$ with respect to time as x_3, the equations of motion are:

$$Mx'' = -K(x_1) - K(x_1 - x_2)$$
$$Mx_2'' = -K(x_2) - K(x_2 - x_1)$$

Since we expect oscillatory motion, we try:

$$x_1(t) = A_1 e^{i\omega t}$$
$$x_2(t) = A_2 e^{i\omega t}$$

Substituting these into the equations of motion gives us:

$$-\omega^2 M A_1 e^{i\omega t} = -2KA_1 e^{i\omega t} + KA_2 e^{i\omega t}$$
$$-\omega^2 M A_2 e^{i\omega t} = KA_1 e^{i\omega t} + 2KA_2 e^{i\omega t}$$

Since the exponential factor is common to all terms, we omit it and simplify:

$$\left(\omega^2 M - 2K\right)A_1 + KA_2 = 0$$
$$KA_1 + \left(\omega^2 M - 2K\right)A_2 = 0$$

And in matrix representation:

$$\begin{bmatrix} \omega^2 M - 2K & K \\ K & \omega^2 M - 2K \end{bmatrix} \begin{pmatrix} A_1 \\ A_2 \end{pmatrix} = 0$$

For this equation to have a non-trivial solution, the matrix on the left must be singular, therefore the determinant of the matrix must be equal to 0, so:

$$\left(\omega^2 M - 2K\right)^2 - K^2 = 0$$

Solving for ω, we have two solutions:

$$\omega_1 = \sqrt{\frac{K}{M}},$$

$$\omega_2 = \sqrt{\frac{3K}{M}},$$

If we substitute ω_1 into the matrix and solve for (A_1, A_2), we get (1, 1). If we substitute ω_2, we get (1, -1). (These vectors are eigenvectors, and the frequencies are eigenvalues.)

The first normal mode is:

$$\begin{pmatrix} x_1(t) \\ x_2(t) \end{pmatrix} = c_1 \begin{pmatrix} 1 \\ 1 \end{pmatrix} \cos(\omega_1 t + \phi_1)$$

and the second normal mode is:

$$\begin{pmatrix} x_1(t) \\ x_2(t) \end{pmatrix} = c_2 \begin{pmatrix} 1 \\ -1 \end{pmatrix} \cos(\omega_2 t + \phi_2)$$

The general solution is a superposition of the normal modes where c_1, c_2, φ_1, and φ_2, are determined by the initial conditions of the problem.

The process demonstrated here can be generalised and formulated using the formalism of Lagrangian mechanics or Hamiltonian mechanics.

Standing Waves

A standing wave is a continuous form of normal mode. In a standing wave, all the space elements (i.e (x,y,z) coordinates) are oscillating in the same frequency and in phase (reaching the equilibrium point together), but each has a different amplitude.

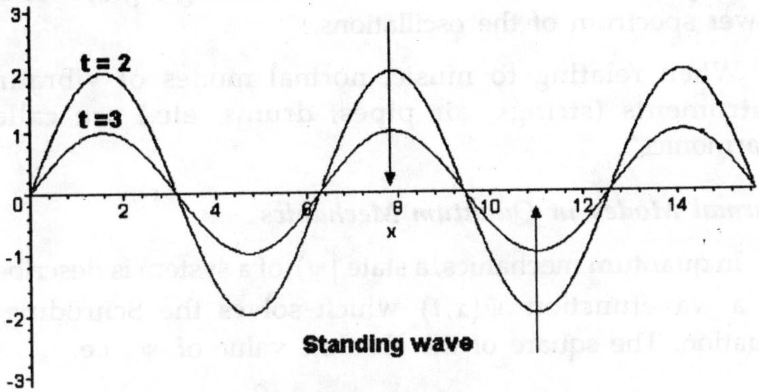

Standing wave

The general form of a standing wave is:

$$\Psi(t) = f(x,y,z)\left(A\cos(\omega t) + B\sin(\omega t)\right)$$

where $f(x, y, z)$ represents the dependence of amplitude on location and the cosine/sine are the oscillations in time.

Physically, standing waves are formed by the interference (superposition) of waves and their reflections (although one may also say the opposite; that a moving wave is a superposition of standing waves). The geometric shape of the medium determines what would be the interference pattern, thus determines the $f(x, y, z)$ form of the standing wave. This space-dependence is called a normal mode.

Usually, for problems with continuous dependence on (x,y,z) there is no single or finite number of normal modes, but there are infinitely many normal modes. If the problem is bounded there are countably many (a discrete infinity of) normal modes (usually numbered $n = 1,2,3,...$). If the problem is not bounded, there is a continuous spectrum of normal modes.

The allowed frequencies are dependent on the normal modes, as well as on physical constants of the problem (density, tension, pressure, etc.) which set the phase velocity of the wave. The range of all possible normal frequencies is called the frequency spectrum. Usually, each frequency is modulated by the amplitude at which it has arisen, creating a graph of the power spectrum of the oscillations.

When relating to music, normal modes of vibrating instruments (strings, air pipes, drums, etc.) are called "harmonics".

Normal Modes in Quantum Mechanics

In quantum mechanics, a state $|\psi\rangle$ of a system is described by a wavefunction $\psi(x,t)$ which solves the Schrödinger equation. The square of the absolute value of ψ, i.e.

$$P(x,t)=|\psi(x,t)|^2$$

is the probability density to measure the particle in place x at time t.

Usually, when involving some sort of potential, the wavefunction is decomposed into a superposition of energy eigenstates, each oscillating with frequency of $\omega=\frac{E_n}{\hbar}$. Thus, we may write

$$|\psi(t)\rangle=\sum_n |n\rangle\langle n|\psi(t=0)\rangle e^{-iE_n t/\hbar}$$

The eigenstates have a physical meaning further than an orthonormal basis. When the energy of the system is measured, the wavefunction collapses into one of its eigenstates and so the particle wavefunction is described by the pure eigenstate corresponding to the measured energy.

Quantum Harmonic Oscillator

The quantum harmonic oscillator is the quantum mechanical analogue of the classical harmonic oscillator. It is one of the most important model systems in quantum mechanics because an arbitrary potential can be approximated as a harmonic potential at the vicinity of a stable equilibrium point.

Furthermore, it is one of the few quantum mechanical systems for which a simple exact solution is known.

Hamiltonian and Energy Eigenstates

In the one-dimensional harmonic oscillator problem, a particle of mass m is subject to a potential $V(x) = (1/2)m\omega^2 x^2$. The Hamiltonian of the particle is:

$$H = \frac{p^2}{2m} + \frac{1}{2}m\omega^2 x^2$$

where x is the position operator, and p is the momentum operator $\left(p = -i\hbar\frac{d}{dx}\right)$. The first term represents the kinetic energy of the particle, and the second term represents the potential energy in which it resides. In order to find the energy levels and the corresponding energy eigenstates, we must solve the time-independent Schrodinger equation,

$$H|\psi\rangle = E|\psi\rangle$$

We can solve the differential equation in the coordinate basis, using a power series method. It turns out that there is a family of solutions,

$$\langle x/\psi_n\rangle = \sqrt{\frac{1}{2^n n!}}.\left(\frac{m\omega}{\pi\hbar}\right)^{1/4}.\exp\left(-\frac{m\omega x^2}{2\hbar}\right).H_n\left(\sqrt{\frac{m\omega}{\hbar}}x\right)$$

n = 0, 1, 2,...

The functions H_n are the Hermite polynomials:

$$H_n(x) = (-1)^n e^{x^2}\frac{d^n}{dx^n}e^{-x^2}$$

They should not be confused with the Hamiltonian, which is also denoted by H. The corresponding energy levels are

$$E_n = \hbar\omega\left(n + \frac{1}{2}\right)$$

This energy spectrum is noteworthy for three reasons. Firstly, the energies are "quantised", and may only take the discrete values of $\hbar\omega$ times 1/2, 3/2, 5/2, and so forth. This is a feature of many quantum mechanical systems. Secondly, the lowest achievable energy is not zero, but $\frac{\hbar\omega}{2}$, which is called the "ground state energy" or zero-point energy. In the ground state, according to quantum mechanics, an oscillator performs null oscillations and its average kinetic energy is positive. It is not obvious that this is significant, because normally the zero of energy is not a physically meaningful quantity, only differences in energies. Nevertheless, the ground state energy has many implications, particularly in quantum gravity. The final reason is that the energy levels are equally spaced, unlike the Bohr model.

Note that the ground state probability density is concentrated at the origin. This means the particle spends most of its time at the bottom of the potential well, as we would expect for a state with little energy. As the energy increases, the probability density becomes concentrated at the "classical turning points", where the state's energy coincides with the potential energy. This is consistent with the classical harmonic oscillator, in which the particle spends most of its time (and is therefore most likely to be found) at the turning points, where it is the slowest. The correspondence principle is thus satisfied.

Ladder Operator Method

The power series solution, though straightforward, is rather tedious. The "ladder operator" method, due to Paul Dirac, allows us to extract the energy eigenvalues without directly solving the differential equation. Furthermore, it is readily generalisable to more complicated problems, notably in quantum field theory. Following this approach, we define the operators a and its adjoint $a^\dagger$

$$a = \sqrt{\frac{m\omega}{2\hbar}}\left(x + \frac{i}{m\omega}p\right)$$

$$a^{\dagger} = \sqrt{\frac{m\omega}{2\hbar}}\left(x - \frac{i}{m\omega}p\right)$$

The operator a is not Hermitian since it and its adjoint $a^{\dagger}$ are not equal.

The operator a and $a^{\dagger}$ have properties as below:

$$a|\phi_n\rangle = \sqrt{n}|\phi_{n-1}\rangle$$

$$a^{\dagger}|\phi_n\rangle = \sqrt{n+1}|\phi_{n+1}\rangle$$

In deriving the form of $a^{\dagger}$, we have used the fact that the operators x and p, which represent observables, *are* Hermitian. These observable operators can be expressed as a linear combination of the ladder operators as

$$x = \sqrt{\frac{\hbar}{2m\omega}}\left(a^{\dagger} + a\right)$$

$$p = i\sqrt{\frac{\hbar m\omega}{2}}\left(a^{\dagger} - a\right)$$

The x and p operators obey the following identity, known as the canonical commutation relation:

$$\left[xp\right] = i\hbar$$

The square brackets in this equation are a commonly-used notational device, known as the commutator, defined as

$$[A,B] \underline{\underline{def}}\ AB - BA$$

Using the above, we can prove the identities

$$H = \hbar\omega\left(a^{\dagger}a + \frac{1}{2}\right)$$

$$\left[a, a^{\dagger}\right] = 1.$$

Now, let $|\psi_E\rangle$ denote an energy eigenstate with energy E. The inner product of any ket with itself must be non-negative, so

$$\left(a|\psi_E\rangle, a|\psi_E\rangle\right) = \langle\psi_E|a^\dagger a|\psi_E\rangle \geq 0$$

Expressing $a^\dagger a$ in terms of the Hamiltonian:

$$\langle\psi_E|\frac{H}{\hbar\omega} - \frac{1}{2}|\psi_E\rangle = \left(\frac{E}{\hbar\omega} - \frac{1}{2}\right) \geq 0,$$

so that $E \geq \frac{\hbar\omega}{2}$. Note that when $\left(a|\psi_E\rangle\right)$ is the zero ket (i.e. a ket with length zero), the inequality is saturated, so that $E = \frac{\hbar\omega}{2}$. It is straightforward to check that there exists a state satisfying this condition; it is the ground (n = 0) state.

Using the above identities, we can now show that the commutation relations of a and $a^\dagger$ with H are:

$$[H, a] = -\hbar\omega a$$

$$\left[H, a^\dagger\right] = -\hbar\omega a^\dagger$$

Thus, provided $\left(a|\psi_E\rangle\right)$ is not the zero ket,

$$\begin{aligned} H\left(a|\psi_E\rangle\right) &= ([H,a] + aH)|\psi_E\rangle \\ &= (-\hbar\omega a + aE)|\psi_E\rangle \\ &= (E - \hbar\omega)\left(a|\psi_E\rangle\right) \end{aligned}$$

Similarly, we can show that

$$H\left(a^\dagger|\psi_E\rangle\right) = (E + \hbar\omega)\left(a^\dagger|\psi_E\rangle\right)$$

In other words, a acts on an eigenstate of energy E to produce, up to a multiplicative constant, another eigenstate of energy $E - \hbar\omega$, and $a^\dagger$ acts on an eigenstate of energy E to produce an eigenstate of energy $(E + \hbar\omega)$. For this reason, a is called a "lowering operator", and $a^\dagger$ a "raising operator". The two operators together are called "ladder operators". In quantum field theory, a and $a^\dagger$ are alternatively called "annihilation" and "creation" operators because they destroy and create particles, which correspond to our quanta of energy.

Given any energy eigenstate, we can act on it with the lowering operator, a, to produce another eigenstate with $\hbar\omega$ - less energy. By repeated application of the lowering operator, it seems that we can produce energy eigenstates down to $E = -\infty$. However, this would contradict our earlier requirement that $E \geq \hbar\omega / 2$. Therefore, there must be a ground-state energy eigenstate, which we label $|0\rangle$ (not to be confused with the zero ket), such that

$$a|0\rangle = 0(zero\,ket).$$

In this case, subsequent applications of the lowering operator will just produce zero kets, instead of additional energy eigenstates. Furthermore, we have shown above that

$$H|0\rangle = \left(\frac{\hbar\omega}{2}\right)|0\rangle$$

Finally, by acting on $|0\rangle$ with the raising operator and multiplying by suitable normalisation factors, we can produce an infinite set of energy eigenstates $\{|0\rangle, |1\rangle, |2\rangle, \ldots, |n\rangle, \ldots\}$, such that

$$H|n\rangle = \hbar\omega\left(n + \frac{1}{2}\right)|n\rangle$$

which matches the energy spectrum.

Natural Length and Energy Scales

The quantum harmonic oscillator possesses natural scales for length and energy, which can be used to simplify the problem. These can be found by non-dimensionalisation. The result is that if we measure energy in units of $\hbar\omega$ and distance in units of $\left(\frac{\hbar}{(m\omega)}\right)^{\frac{1}{2}}$, then the Schrödinger equation becomes:

$$H = -\frac{1}{2}\frac{d^2}{du^2} + \frac{1}{2}u^2,$$

and the energy eigenfunctions and eigenvalues become

$$\langle x \mid \psi_n \rangle = \frac{1}{\sqrt{2^n n!}}\pi^{-1/4}\exp\left(-u^2/2\right)H_n(u)$$

$$E_n = n + \frac{1}{2}.$$

Example: Diatomic Molecules

In diatomic molecules, the natural frequency can be found by:

$$\omega = \sqrt{\frac{k}{m^{\tau}}}$$

where

- $\omega = 2\pi f$ is the angular frequency,
- k is the bond force constant, and
- m_r is the reduced mass.

N-dimensional Harmonic Oscillator

The one-dimensional harmonic oscillator is readily generalisable to N dimensions, where $N = 1, 2, 3, \ldots$. In one dimension, the position of the particle was specified by a single coordinate, x. In N dimensions, this is replaced by N position coordinates, which we label $x_1, \ldots, x_N$. Corresponding to each position coordinate is a momentum; we label these $p_1, \ldots, p_N$.

The canonical commutation relations between these operators are

$$\begin{aligned} \left[x_i, p_j\right] &= i\hbar\delta_{i,j} \\ \left[x_i, x_j\right] &= 0 \\ \left[p_i, p_j\right] &= 0 \end{aligned}.$$

The Hamiltonian for this system is

$$H = \sum_{i=1}^{N}\left(\frac{p_i^2}{2m} + \frac{1}{2}m\omega^2 x_i^2\right).$$

As the form of this Hamiltonian makes clear, the N-dimensional harmonic oscillator is exactly analogous to N independent one-dimensional harmonic oscillators with the same mass and spring constant. In this case, the quantities $x_1, \ldots, x_N$ would refer to the positions of each of the N particles. This is a happy property of the r^2 potential, which allows the potential energy to be separated into terms depending on one coordinate each.

This observation makes the solution straightforward. For a particular set of quantum numbers $\{n\}$, the energy eigenfunctions for the N-dimensional oscillator are expressed in terms of the 1-dimensional eigenfunctions as:

$$\left\langle x \mid \psi_{\{n\}} \right\rangle = \prod_{i=1}^{N}\left\langle x_i \mid \psi_{n_i} \right\rangle$$

In the ladder operator method, we define N sets of ladder operators,

$$a_i = \sqrt{\frac{m\omega}{2\hbar}}\left(x_i + \frac{i}{m\omega}pi\right)$$

$$a_i^{\dagger} = \sqrt{\frac{m\omega}{2\hbar}}\left(x_i - \frac{i}{m\omega}pi\right)$$

By a procedure analogous to the one-dimensional case, we can then show that each of the a_i and $a_i^\dagger$ operators lower and raise the energy by $\hbar\omega$ respectively. The Hamiltonian is

$$H = \hbar\omega \sum_{i=1}^{N} \left(a_i^\dagger a_i + \frac{1}{2} \right)$$

This Hamiltonian is invariant under the dynamic symmetry group *U(N)* (the unitary group in *N* dimensions), defined by

$$U a_i^\dagger U^\dagger = \sum_{j=1}^{N} a_i^\dagger U_{ji} \quad \text{for all } U \in U(N)$$

where U_{ji} is an element in the defining matrix representation of *U(N)*.

The energy levels of the system are

$$E = \hbar\omega \left[(n_i + \ldots + n_N) + \frac{N}{2} \right]$$

$n_i = 0, 1, 2, \ldots$ (the number of bosons in mode i).

As in the one-dimensional case, the energy is quantised. The ground state energy is *N* times the one-dimensional energy, as we would expect using the analogy to *N* independent one-dimensional oscillators. There is one further difference: in the one-dimensional case, each energy level corresponds to a unique quantum state. In *N*-dimensions, except for the ground state, the energy levels are *degenerate,* meaning there are several states with the same energy.

The degeneracy can be calculated relatively easily, as an example, consider the 3-dimensional case: Define $n = n_1 + n_2 + n_3$. All states with the same n will have the same energy. For a given n, we choose a particular n_1. Then $n_2 + n_3 = n - n_1$. There are $n - n_1 + 1$ possible groups $\{n_2, n_3\}$. n_2 can take on the values 0 to $n - n_1$, and for each n_2 the value of n_3 is fixed. The degree of degeneracy therefore is:

$$g_n = \sum_{n_1=0}^{n} n - n_1 + 1 = \frac{(n+1)(n+2)}{2}$$

Formula for general *N* and *n* [g_n being the dimension of the symmetric irreducible n^{th} power representation of the unitary group *U(N)*]:

$$g_n = \binom{N+n-1}{n}$$

The special case *N* = 3, given above, follows directly from this general equation.

Example: 3D Isotropic Harmonic Oscillator

The Schrödinger equation of a spherically-symmetric three-dimensional harmonic oscillator can be solved explicitly by separation of variables. This procedure is analogous to the separation performed in the hydrogen-like atom problem, but with the spherically symmetric potential

$$V(r) = \frac{1}{2}\mu\omega^2 r^2,$$

where μ is the mass of the problem. Because *m* will be used below for the magnetic quantum number, mass is indicated by μ, instead of *m*, as earlier in this article.

The solution reads

$$\psi_{klm}(r,\theta,\phi) = N_{kl} r^l e^{-vr^2} L_k^{\left(l+\frac{1}{2}\right)}\left(2Vr^2\right) Y_{lm}(\theta,\phi)$$

where

N_{kl} is a normalisation constant.

$$V \equiv \frac{\mu\omega}{2\hbar}$$

$L_k^{\left(l+\frac{1}{2}\right)}\left(2Vr^2\right)$ are generalised Laguerre polynomials. The order *k* of the polynomial is a non-negative integer.

$Y_{lm}(\theta,\phi)$ is a spherical harmonic function.

$\hbar$ is the reduced Planck constant: $\hbar \equiv \dfrac{h}{2\pi}$.

The energy eigenvalue is

$$E = \hbar\omega\left(2k + l + \frac{3}{2}\right)$$

The energy is usually described by the single quantum number

$$n \equiv 2k + l$$

Because k is a non-negative integer, for every even n we have $l = 0,2...,n$ " $2,n$ and for every odd n we have $l = 1,3...,n$ " $2,n$. The magnetic quantum number m is an integer satisfying $-l \le m \le l$, so for every n and l there are $2l+1$ different quantum states, labelled by m. Thus, the degeneracy at level n is

$$\sum_{l=...,n-2,n} (2l+1) = \frac{(n+1)(n+2)}{2}$$

where the sum starts from 0 or 1, according to whether n is even or odd. This result is in accordance with the dimension formula above.

Coupled Harmonic Oscillators

In this problem, we consider N equal masses which are connected to their neighbours by springs, in the limit of large N. The masses form a linear chain in one dimension, or a regular lattice in two or three dimensions.

We denote the positions of the masses by x_1, x_2, ..., as measured from their equilibrium positions (i.e. $x_k = 0$ if particle k is at its equilibrium position.) In two or more dimensions, the xs are vector quantities. The Hamiltonian of the total system is

$$H = \sum_{i=1}^{N} \frac{p_i^2}{2m} + \frac{1}{2} m\omega^2 \sum_{\{ij\}(nn)} \left(x_i - x_j\right)^2$$

The potential energy is summed over "nearest-neighbour" pairs, so there is one term for each spring.

Remarkably, there exists a coordinate transformation to turn this problem into a set of independent harmonic oscillators, each of which corresponds to a particular collective distortion of the lattice. These distortions display some particle-like properties, and are called phonons. Phonons occur in the ionic lattices of many solids, and are extremely important for understanding many of the phenomena studied in solid state physics.

Anharmonicity

Anharmonicity is the deviation of a system from being a harmonic oscillator. An oscillator that is not oscillating in simple harmonic motion is known as an anharmonic oscillator where the system can be approximated to a harmonic oscillator and the anharmonicity can be calculated using perturbation theory. If the anharmonicity is large then other numerical techniques have to be used.

Anharmonic Ratio (cross-ratio)

If *A, B, C, D* all lie on a straight line then the anharmonic ratio is:

$$(A,B,C,D) = \frac{AC/AD}{BC/BD}$$

In the complex plane the anharmonic ratio is

$$(z_1, z_2; z_3, z_4) = \frac{(z_1 - z_3)(z_2 - z_4)}{(z_1 - z_4)(z_2 - z_3)}$$

- There are six distinct anharmonic ratios depending on the order of which the points are chosen.
- Anharmonicity plays a role in lattice and molecular vibrations, in quantum oscillations, and in Acoustics.

Parametric Oscillator

A parametric oscillator is a simple harmonic oscillator whose parameters (its resonant frequency ω and damping β) vary in time in a defined way

$$\frac{d^2x}{dt^2} + \beta(t)\frac{dx}{dt} + w^2(t)x = 0$$

This equation is linear in x(t). By assumption, the parameters ω^2 and β depend only on time and do not depend on the state of the oscillator. In general, β (t) and/or ω^2 (t) are assumed to vary periodically with the same period T.

Remarkably, if the parameters vary at roughly twice the natural frequency of the oscillator, the oscillator phase-locks to the parametric variation and absorbs energy at a rate proportional to the energy it already has. Without a compensating energy-loss mechanism, the oscillation amplitude grows exponentially. (This phenomenon is called parametric excitation, parametric resonance or parametric pumping.) However, if the initial amplitude is zero, it will remain so; this distinguishes it from the non-parametric resonance of driven simple harmonic oscillators, in which the amplitude grows linearly in time regardless of the initial state.

A familiar experience of parametric oscillation is playing on a swing. By alternately raising and lowering their centre of mass (changing their moment of inertia and, thus, the resonant frequency) at key points in the swing, children can quickly reach large amplitudes provided that they have some amplitude to start with (e.g., get a push). Doing so at rest, however, goes nowhere.

Transformation of the Equation

We begin by making a change of variables

$$q(t) \underline{\underline{\mathrm{def}}}\ e^{D(t)} x(t)$$

where $D(t)$ is a time integral of the damping

$$D(t) \underline{\underline{\mathrm{def}}} \frac{1}{2} \int^{t} d\tau\, \beta(\tau).$$

This change of variables eliminates the damping term

$$\frac{d^2 q}{dt^2} + \Omega^2(t) q = 0$$

where the transformed frequency is defined

$$\Omega^2(t) = \omega^2(t) - \frac{1}{2}\left(\frac{d\beta}{dt}\right) - \frac{1}{4}\beta^2.$$

In general, the variations in damping and frequency are relatively small perturbations

$$\beta(t) = \omega_0\left[b + g(t)\right]$$

$$\omega^2(t) = \omega_0^2\left[1 + h(t)\right]$$

where ω_0 and $b\omega_0$ are constants, namely, the time-averaged oscillator frequency and damping, respectively. The transformed frequency can be written in a similar way:

$$\Omega^2(t) = \omega_0^2\left[1 + f(t)\right],$$

where ω_n is the natural frequency of the damped harmonic oscillator

$$\omega_0^2 \underline{\underline{\text{def}}}\, \omega_0^2\left(1 - \frac{b^2}{4}\right)$$

and

$$\omega_n^2 f(t) \underline{\underline{\text{def}}}\, \omega_0^2 h(t) - \frac{1}{2\omega_0}\left(\frac{dg}{dt}\right) - \frac{b}{2}g(t) - \frac{1}{4}g^2(t).$$

Thus, our transformed equation can be written

$$\frac{d^2q}{dt^2} + \omega_n^2\left[1 + f(t)\right]q = 0.$$

Remarkably, the independent variations $g(t)$ and $h(t)$ in the oscillator damping and resonant frequency, respectively, can be combined into a single pumping function $f(t)$. The converse conclusion is that any form of parametric excitation can be accomplished by varying either the resonant frequency or the damping, or both.

Solution of the Transformed Equation

Let us assume that $f(t)$ is sinusoidal, specifically

$$f(t) = f_0 \sin 2\omega_p t$$

where the pumping frequency $2\omega_p \approx 2\omega_n$ but need not equal $2\omega_n$ exactly. The solution $q(t)$ of our transformed equation may be written

$$q(t) = A(t)\cos \omega_p t + B(t)\sin \omega_p t$$

where we have factored out the rapidly varying components ($\cos \omega_p t$ and $\sin \omega_p t$) to isolate the slowly varying amplitudes $A(t)$ and $B(t)$. This corresponds to Laplace's variation of parameters method.

Substituting this solution into the transformed equation and retaining only the terms first-order in $f_0 \ll 1$ yields two coupled equations

$$2\omega_p \frac{dA}{dt} = \left(\frac{f_0}{2}\right)\omega_n^2 A - \left(\omega_p^2 - \omega_n^2\right)B$$

$$2\omega_p \frac{dB}{dt} = -\left(\frac{f_0}{2}\right)\omega_n^2 B + \left(\omega_p^2 - \omega_n^2\right)A$$

We may decouple and solve these equations by making another change of variables

$$A(t) \underline{\underline{\text{def}}}\, r(t)\cos\theta(t)$$

$$B(t) \underline{\underline{\text{def}}}\, r(t)\sin\theta(t)$$

which yields the equations

$$\frac{dr}{dt} = (\alpha_{\max} \cos 2\theta) r$$

$$\frac{d\theta}{dt} = -\alpha_{\max}\left[\sin 2\theta - \sin 2\theta_{eq}\right]$$

where we have defined for brevity

$$\alpha_{\max} \underline{\underline{\text{def}}} \frac{f_0 \omega_n^2}{4\omega_p}$$

$$\sin 2\theta_{eq} \underline{\underline{\text{def}}} \left(\frac{2}{f_0}\right) \in$$

and the detuning

$$\epsilon \underline{\underline{\text{def}}} \frac{\omega_p^2 - \omega_n^2}{\omega_n^2}$$

The θ equation does not depend on r, and linearisation near its equilibrium position θ_{eq} shows that θ decays exponentially to its equilibrium—

$$\theta(t) = \theta_{eq} + \left(\theta_0 - \theta_{eq}\right) e^{-2\alpha t}$$

where the decay constant

$$\alpha \underline{\underline{\text{def}}}\, \alpha_{max} \cos 2\theta_{eq}.$$

In other words, the parametric oscillator phase-locks to the pumping signal $f(t)$.

Taking $\theta(t) = \theta_{eq}$ (i.e., assuming that the phase has locked), the r equation becomes

$$\frac{dr}{dt} = \alpha r$$

whose solution is $r(t) = r_0 e^{\alpha t}$; the amplitude of the $q(t)$ oscillation diverges exponentially. However, the corresponding amplitude $R(t)$ of the untransformed $x \underline{\underline{\text{def}}}\, q e^{-D}$ variable need not diverge.

$$R(t) = r(t) e^{"D} = r_0 e^{\alpha t - D}$$

The amplitude $R(t)$ diverges, decays or stays constant, depending on whether αt is greater than, less than, or equal to D, respectively.

The maximum growth rate of the amplitude occurs when $\omega_p = \omega_n$. At that frequency, the equilibrium phase θ_{eq} is zero, implying that $\cos 2\theta_{eq} = 1$ and $\alpha = \alpha_{max}$. As ω_p is varied from ω_n, θ_{eq} moves away from zero and $\alpha < \alpha_{max}$, i.e., the amplitude grows more slowly. For sufficiently large deviations of ω_p, the decay constant á can become purely imaginary since

$$\alpha = \alpha_{max} \sqrt{1 - \left(\frac{2}{f_0}\right)^2 \epsilon^2}$$

If the detuning ε exceeds $f_0/2$, α becomes purely imaginary and $q(t)$ varies sinusoidally. Using the definition of the detuning ε, the pumping frequency $2\omega_p$ must lie between $2\omega_n\sqrt{1-\frac{f_0}{2}}$ and $2\omega_n\sqrt{1+\frac{f_0}{2}}$. Expanding the square roots in a binomial series shows that the spread in pumping frequencies that result in exponentially growing q is approximately $\omega_n f_0$.

Intuitive Understanding of Parametric Excitation

The above derivation may seem like a mathematical sleight-of-hand, so it may be helpful to give an intuitive derivation. The q equation may be written in the form

$$\frac{d^2q}{dt^2}+\omega_n^2 q=-\omega_n^2 f(t)q$$

which represents a simple harmonic oscillator (or, alternatively, a bandpass filter) being driven by a signal $-\omega_n^2 f(t)q$ that is proportional to its response q.

Assume that $q(t) = A\cos\omega_p t$ already has an oscillation at frequency ω_p and that the pumping $f(t) = f_0\sin 2\omega_p t$ has double the frequency and a small amplitude $f_0 \ll 1$. Their product $q(t)f(t)$ produces two driving signals, one at frequency ω_p and the other at frequency $3\omega_p$

$$f(t)q(t)=\frac{f_0}{2}A\left(\sin\omega_p t+\sin 3\omega_p t\right)$$

Being off-resonance, the $3\omega_p$ signal is attentuated and can be neglected initially. By contrast, the ω_p signal is on resonance, serves to amplify q and is proportional to the amplitude A. Hence, the amplitude of q grows exponentially unless it is initially zero.

Expressed in Fourier space, the multiplication $f(t)q(t)$ is a convolution of their Fourier transforms $\tilde{F}(\omega)$ and $\tilde{Q}(\omega)$. The positive feedback arises because the + 2 ω_p component of $f(t)$ converts the $-\omega_p$ component of $q(t)$ into a driving signal at $+\omega_p$, and vice versa (reverse the signs). This explains why the pumping frequency must be near 2 ω_n, twice the natural frequency of the oscillator. Pumping at a grossly different frequency would not couple (i.e., provide mutual positive feedback) between the " ω_p and $+\omega_p$ components of $q(t)$.

Parametric Amplifiers

The parametric oscillator equation can be extended by adding an external driving force $E(t)$:

$$\frac{d^2x}{dt^2}+\beta(t)\frac{dx}{dt}+\omega^2(t)x=E(t)$$

We assume that the damping D is sufficiently strong that, in the absence of the driving force E, the amplitude of the parametric oscillations does not diverge, i.e., that $\alpha\, t < D$. In this situation, the parametric pumping acts to lower the effective damping in the system. For illustration, let the damping be constant $\beta(t) = \omega_0 b$ and assume that the external driving force is at the mean resonant frequency ω_0, i.e., $E(t) = E_0 \sin\omega_0 t$. The equation becomes

$$\frac{d^2x}{dt^2}+b\omega_0\frac{dx}{dt}+\omega_0^2\left[1+h_0\sin 2\omega_0 t\right]x=E_0\sin\omega_0 t$$

whose solution is roughly

$$x(t)=\frac{2E_0}{\omega_0(2b-h_0)}\cos\omega_0 t$$

As h_0 approaches the threshold $2b$, the amplitude diverges. When $h \geq 2b$, the system enters parametric resonance and the amplitude begins to grow exponentially, even in the absence of a driving force $E(t)$.

Other Relevant Mathematical Results

If the parameters of any second-order linear differential

equation are varied periodically, Floquet analysis shows that the solutions must vary either sinusoidally or exponentially.

The q equation above with periodically varying $f(t)$ is an example of a Hill equation. If $f(t)$ is a simple sinusoid, the equation is called a Mathieu equation.

History

Faraday (1831) was the first to notice oscillations of one frequency being excited by forces of double the frequency, in the crispations (ruffled surface waves) observed in a wine glass excited to "sing". Melde (1859) generated parametric oscillations in a string by employing a tuning fork to periodically vary the tension at twice the resonant frequency of the string. Parametric oscillation was first treated as a general phenomenon by Rayleigh (1883,1887), whose papers are still worth reading today.

Parametric oscillators have been developed as low-noise amplifiers, especially in the radio and microwave frequency range. Thermal noise is minimal, since a reactance (not a resistance) is varied. Another common use is frequency conversion, e.g., conversion from audio to radio frequencies. Parametric amplifiers (paramps) were first used in 1913-1915 for radio telephony from Berlin to Vienna and Moscow, and were predicted to have a useful future (Alexanderson, 1916). The early paramps varied inductances, but other methods have been developed since, e.g., the varactor diodes, klystron tubes, Josephson junctions and optical methods. A practical parametric oscillator needs two three connections. One for common, one to feed the pump, and one to retrieve the oscillation and maybe a fourth one for biasing. A parametric amplifier needs a fifth port for the seed. Since a varactor diode has only two connections it can only be a part of a LC network with four eigenvectors with nodes at the connections. This can be implemented as a transimpedance amplifier, a travelling wave amplifier or by means of a circulator.

5

Different Electronic Oscillators

An electronic oscillator is an electronic circuit that produces a repetitive electronic signal, often a sine wave or a square wave.

A low-frequency oscillator (LFO) is an electronic oscillator that generates an AC waveform between 0.1 Hz and 10 Hz. This term is typically used in the field of audio synthesisers, to distinguish it from an audio frequency oscillator.

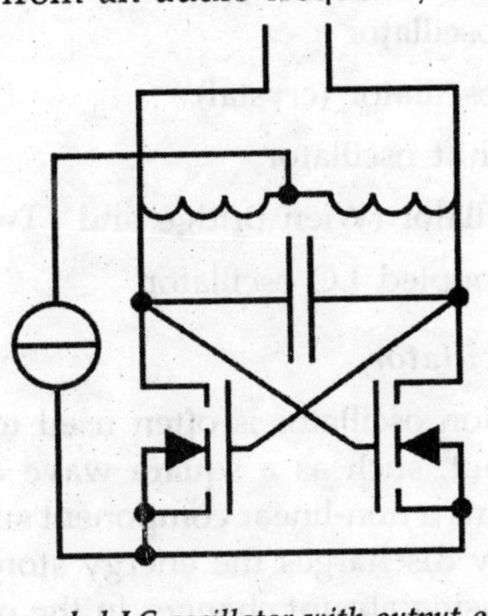

Cross coupled LC oscillator with output on top

Types of Electronic Oscillator

There are two main types of electronic oscillator: the harmonic oscillator and the relaxation oscillator.

Harmonic Oscillator

The harmonic oscillator produces a sinusoidal output. The basic form of a harmonic oscillator is an electronic amplifier with the output attached to a narrow-band electronic filter, and the output of the filter attached to the input of the amplifier. When the power supply to the amplifier is first switched on, the amplifier's output consists only of noise. The noise travels around the loop, being filtered and re-amplified until it increasingly resembles the desired signal.

A piezoelectric crystal (commonly quartz) may be coupled to the filter to stabilise the frequency of oscillation, resulting in a crystal oscillator.

There are many ways to implement harmonic oscillators, because there are different ways to amplify and filter. For example:

- Hartley oscillator
- Colpitts oscillator
- Clapp oscillator
- Pierce oscillator (crystal)
- Phase-shift oscillator
- RC oscillator (Wien Bridge and "Twin-T")
- Cross-coupled LC oscillator

Relaxation Oscillator

The relaxation oscillator is often used to produce a non-sinusoidal output, such as a square wave or sawtooth. The oscillator contains a non-linear component such as a transistor that periodically discharges the energy stored in a capacitor or inductor, causing abrupt changes in the output waveform.

Square-wave relaxation oscillators can be used to provide the clock signal for sequential logic circuits such as timers and counters, although crystal oscillators are often preferred for their greater stability.

Triangle-wave or sawtooth oscillators are used in the timebase circuits that generate the horizontal deflection signals for cathode ray tubes in analogue oscilloscopes and television sets. In function generators, this triangle wave may then be further shaped into a close approximation of a sine wave. Other types of relaxation oscillators include the multivibrator and the rotary travelling wave oscillator.

Analog Temperature Controlled Crystal Oscillator

In physics, an Analog Temperature Controlled Crystal Oscillator or Analogue Temperature Compensated Crystal Oscillator (ATCXO) uses analog sampling techniques to correct the temperature deficiencies of a crystal oscillator circuit, its package and its environment.

Typically the correction techniques involve the physical and electrical characterisation of the motional inductance and terminal capacitance of a crystal blank, the knowledge of which is used to create a correction polynomial, or algorithm, which in turn is implemented in circuit blocks by Electronic Design Engineers. These circuit blocks are usually simulated in a mathematical modelling software tool such as SPICE, to verify that the original measured data can be corrected adequately. Once the system performance has been verified, these circuits are then implemented in a silicon die, usually in a bulk CMOS technology. Once fabricated, this die is then embedded into an oscillator module along with the crystal blank. Due to the sub 1ppm accuracy of this type of crystal oscillator specialist packaging must be used to ensure good ageing and temperature shock characteristics Example applications are for use in low power or battery operated consumer electronic products such as GSM or CDMA mobile phones, or GPS satellite navigation systems.

Armstrong (tickler) Oscillator

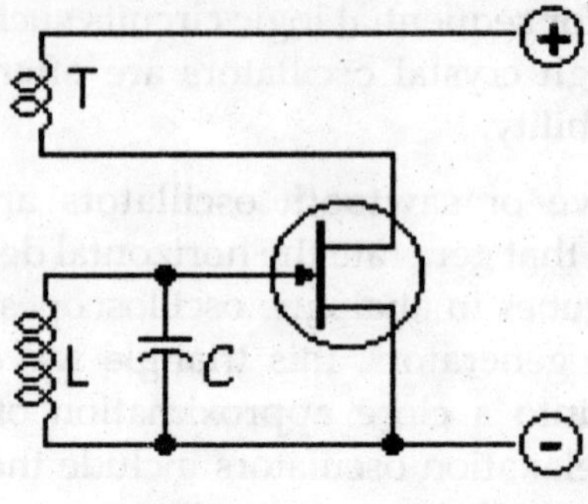

A basic Armstrong oscillator

The Armstrong oscillator is an oscillator used to produce a sine-wave output of constant amplitude and of fairly constant frequency within the RF range. It is generally used as a local oscillator in receivers, as a source in signal generators, and as a radio-frequency oscillator in the medium- and high-frequency range. The oscillator was name after its inventor Edwin Armstrong.

Autogenerator

Autogenerator - the single transistor (in the older devices — vacuum tube) generator that was commonly used as the local oscillator in the superheterodyne radio receivers.

The autogenerator frequency is determined by the oscillatory LC loop with the variable capacitor (hence the frequency of the autogenerator can be tuned).

The Work of Transistor

The typical autogenerator consumes 1.5-4 mA of current (I). During the normal work of transistor, the voltage between the ground and emitter, U_e, is equal to about 0.2 of the voltage of the power source (E_n), and the voltage between the ground and collector is equal to about 0.8 of the power source. This mode is determined by resistors R1 - R4 that are also responsible for the thermal stability of the generator. The oscillatory loop, determining the frequency of the oscillation, is formed by the L_k and C_4.

The Reasons of Choosing between Designs

Autogenerators, producing relatively low frequencies (below 2-4 MHz) usually use additional windings on the oscillatory LC loop to provide the positive feedback, required for the generation. The oscillatory loop in this case also works like transformer or autotransformer. For the higher frequencies the computed number of windings is too small (surely cannot be less than one), and it is easier to create the positive feedback using the capacitory divider.

Beat Frequency Oscillator

A beat frequency oscillator or BFO in radio telegraphy, is a dedicated oscillator used to create an audio frequency signal from carrier wave transmissions to make them audible, as they are not broadcast as such. The signal from the BFO is then heterodyned with the intermediate frequency signal to create an audio frequency signal.

Example: A receiver is tuned to a Morse code signal, and the receiver's intermediate frequency (IF) is F_{if} = 45000 Hz. That means the dots and dashes [also known as dits and dahs] have become pulses of a 45000 Hz tone, which is inaudible. To make them audible, the frequency needs to be shifted into the audio range, for instance $F_{baseband}$ = 1000 Hz. To do that, the desired frequency shift is F_{bfo} = 44000 Hz, and the desired BFO output waveform is *cos(2ð 44000 t)*. When the signal at frequency F_{if} is multiplied by that waveform, it shifts the signal to two other frequencies: $|F_{if}$ " $F_{bfo}|$ and $(F_{if} + F_{bfo})$. The *difference frequency,* $|F_{if}$ " $F_{bfo}|$ = 1000 Hz, is also known as the beat frequency. The other frequency, $(F_{if} + F_{bfo})$ = 89000 Hz, can then be removed by a lowpass filter (such as an ordinary speaker (The speaker cannot vibrate at such a high frequency) or the humans ear (The frequency of the sound is outside the humans hearing range).

Blocking Oscillator

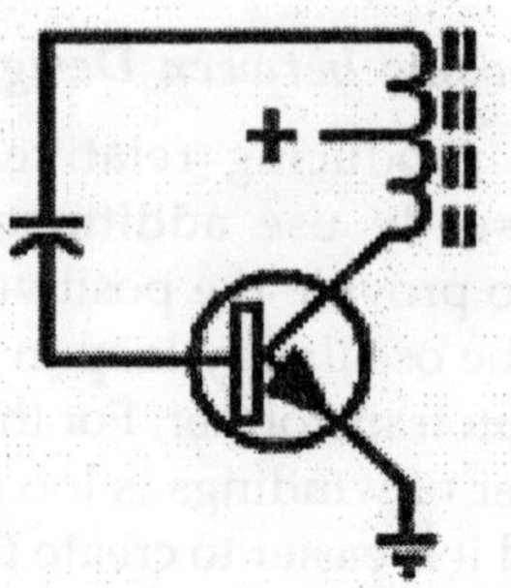

Solid-state Blocking Oscillator (using NPN transistor)

A blocking oscillator is the minimal configuration of discrete electronic components which can produce a free-running signal, requiring only a capacitor, transformer, and one amplifying component. The name is derived from the fact that the transistor (or tube) is cutoff or "blocked" for most of the duty-cycle, producing periodic pulses. The non-sinusoidal output is not suitable for use as a radio-frequency local oscillator, but it can serve to flash lights or LEDs, and the simple tones are sufficient for applications such as alarms or a morse-code practice device. Some cameras use a blocking oscillator to strobe the flash prior to a shot to reduce the red-eye effect.

Due to the simplicity of the circuit, it forms the basis for many of the learning projects in commercial electronic kits. A secondary winding of the transformer can be fed to a speaker, a lamp, or the windings of a relay. A potentiometer placed in parallel with the timing capacitor permits the frequency to be adjusted, but at low resistances the transistor will be overdriven, and possibly damaged. The output signal will jump in amplitude and be greatly distorted. The frequency of the oscillator is also affected by the supply voltage.

Clapp Oscillator

The Clapp oscillator is one of several types of electronic oscillator constructed from a transistor (or vacuum tube) and a positive feedback network. It was invented by James K. Clapp in 1948.

Referring the notional circuit in the figure, the network is comprised of a single inductor and three capacitors, with two capacitors (C_1 and C_2) forming a voltage divider that determines the amount of feedback voltage applied to the transistor input. The Clapp oscillator is a Colpitts oscillator with an additional capacitor placed in series with the inductor. The oscillation frequency in hertz (cycles per second) for the circuit in the figure, which uses a field-effect transistor (FET), is

$$f_0 = \frac{1}{2\pi}\sqrt{\frac{1}{L}\left(\frac{1}{C_0}+\frac{1}{C_1}+\frac{1}{C_2}\right)}$$

A Clapp circuit is often preferred over a Colpitts circuit for constructing a variable frequency oscillator (VFO) that uses a single variable capacitor to adjust the frequency. In a Colpitts VFO, the voltage divider contains the variable capacitor (either C_1 or C_2). This causes the feedback voltage to be variable as well, sometimes making the Colpitts circuit less likely to achieve oscillation over a portion of the desired frequency range. This problem is avoided in the Clapp circuit by using fixed capacitors in the voltage divider and a variable capacitor (C_0) in series with the inductor.

Colpitts Oscillator

A Colpitts oscillator, named after its inventor Edwin H. Colpitts, is one of a number of designs for electronic oscillator circuits. One of the key features of this type of oscillator is its simplicity and robustness. It is not difficult to achieve satisfactory results with little effort.

A Colpitts oscillator is the electrical dual of a Hartley oscillator. In the Colpitts circuit, two capacitors and one inductor determine the frequency of oscillation. The Hartley circuit uses two inductors (or a tapped single inductor) and one capacitor. The capacitor can be a variable device by using a varactor.

The following schematic is an example using an NPN transistor in the common base configuration. Frequency of oscillation is roughly 50 MHz:

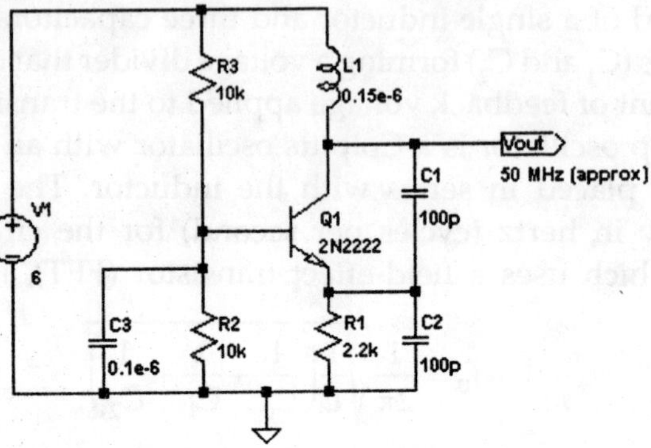

The following schematic is a common collector version of the same oscillator:

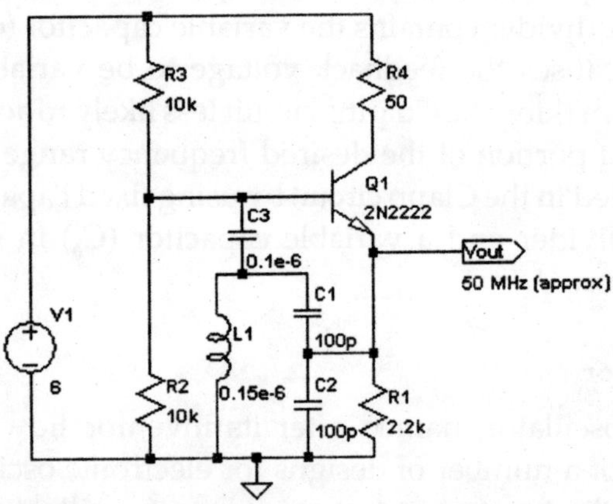

The bipolar transistor could be replaced with a JFET or other active device capable of producing gain at the oscillation frequency.

The ideal frequency of oscillation is given by this equation:

$$f_0 = \frac{1}{2\pi\sqrt{L_1 \cdot \left(\frac{C_1 . C_2}{C_1 + C_2}\right)}}$$

where the series combination of C_1 and C_2 creates the effective load capacitance of the LC tank.

Given the values in the circuit above, the equation predicts a frequency of roughly 58 MHz. The real circuit will oscillate at a slightly lower frequency due to junction capacitances of the transistor and possibly other stray capacitances.

Analysis

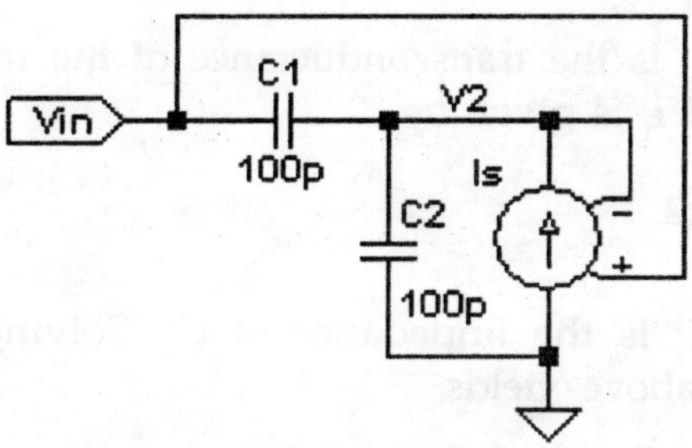

Colpitts oscillator model

One method of oscillator analysis is to determine the input impedance of an input port neglecting any reactive components. If the impedance yields a negative resistance term, oscillation is possible. This method will be used here to determine conditions of oscillation and the frequency of oscillation.

For initial analysis, parasitic elements and device non-linearities will be ignored. These terms can be included later in a more rigorous analysis. Even with these approximations, acceptable comparison with experimental results is possible.

Ignoring the inductor, the input impedance can be written as:

$$Z_{in} = \frac{v_1}{i_1}$$

Where v_1 is the input voltage and i_1 is the input current. The voltage v_2 is given by

$v_2 = i_2 Z_2$

Where Z_2 is the impedance of C_2. The current flowing into C_2 is i_2, which is the sum of two currents:

$i_2 = i_1 + i_s$

Where i_s is the current supplied by the transistor. i_s is a dependent current source given by

$i_s = g_m (v_1 - v_2)$

Where g_m is the transconductance of the transistor. The input current i_1 is given by

$$i_1 = \frac{v_1 - v_2}{Z_1}$$

Where Z_1 is the impedance of C_1. Solving for v_2 and substituting above yields

$Z_{in} = Z_1 + Z_2 + g_m Z_1 Z_2$

The input impedance appears as the two capacitors in series with an interesting term, R_{in} which is proportional to the product of the two impedances:

$R_{in} = g_m \cdot Z_1 \cdot Z_2$

If Z_1 and Z_2 are complex and of the same sign, R_{in} will be a negative resistance. If the impedances for Z_1 and Z_2 are substituted, R_{in} is

$$R_{in} = \frac{-g\,m}{\omega^2 C_1 C_2}$$

If an inductor is connected to the input, the circuit will oscillate if the magnitude of the negative resistance is greater than the resistance of the inductor and any stray elements.

For the example oscillator above, the emitter current is roughly 1 mA. The transconductance is roughly 40 mS. Given all other values, the input resistance is roughly

$R_{in} = -3\,\Omega$

This value should be sufficient to overcome any positive resistance in the circuit. By inspection, oscillation is more likely for larger values of transconductance and/or smaller values of capacitance. A more complicated analysis of the common-base oscillator reveals that a low frequency amplifier voltage gain must be at least four to achieve oscillation. The low frequency gain is given by:

$$A_v = g_m \cdot R_p \geq 4$$

If the two capacitors are replaced by inductors and magnetic coupling is ignored, the circuit becomes a Hartley oscillator. In that case, the input impedance is the sum of the two inductors and a negative resistance given by:

$$R_{in} = -g_m \omega^2 L_1 L_2$$

In the Hartley circuit, oscillation is more likely for larger values of transconductance and/or larger values of inductance.

Crystal Oscillator

A crystal oscillator is an electronic circuit that uses the mechanical resonance of a vibrating crystal of piezoelectric material to create an electrical signal with a very precise frequency. This frequency is commonly used to keep track of time (as in quartz wristwatches), to provide a stable clock signal for digital integrated circuits, and to stabilise frequencies for radio transmitters.

Using an amplifier and feedback, it is an especially accurate form of an electronic oscillator. The crystal used therein is sometimes called a "timing crystal".

Crystals for Timing Purposes

A crystal is a solid in which the constituent atoms, molecules, or ions are packed in a regularly ordered, repeating pattern extending in all three spatial dimensions.

Almost any object made of an elastic material could be

used like a crystal, with appropriate transducers, since all objects have natural resonant frequencies of vibration. For example, steel is very elastic and has a high speed of sound. It was often used in mechanical filters before quartz. The resonant frequency depends on size, shape, elasticity, and the speed of sound in the material. High-frequency crystals are typically cut in the shape of a simple, rectangular plate. Low-frequency crystals, such as those used in digital watches, are typically cut in the shape of a tuning fork. For applications not needing very precise timing, a low-cost ceramic resonator is often used in place of a quartz crystal.

When a crystal of quartz is properly cut and mounted, it can be made to distort in an electric field by applying a voltage to an electrode near or on the crystal. This property is known as piezoelectricity. When the field is removed, the quartz will generate an electric field as it returns to its previous shape, and this can generate a voltage. The result is that a quartz crystal behaves like a circuit composed of an inductor, capacitor and resistor, with a precise resonant frequency.

Quartz has the further advantage that its elastic constants and its size change in such a way that the frequency dependence on temperature can be very low. The specific characteristics will depend on the mode of vibration and the angle at which the quartz is cut (relative to its crystallographic axes). Therefore, the resonant frequency of the plate, which depends on its size, will not change much, either. This means that a quartz clock, filter or oscillator will remain accurate. For critical applications the quartz oscillator is mounted in a temperature-controlled container, called a crystal oven, and can also be mounted on shock absorbers to prevent perturbation by external mechanical vibrations.

Quartz timing crystals are manufactured for frequencies from a few tens of kilohertz to tens of megahertz. More than two billion (2×10^9) crystals are manufactured annually. Most are small devices for consumer devices such as wristwatches, clocks, radios, computers, and cellphones. Quartz crystals are also found inside test and measurement equipment, such as counters, signal generators, and oscilloscopes.

Crystal Modelling

A quartz crystal can be modelled as an electrical network with a low impedance (series) and a high impedance (parallel) resonance point spaced closely together. Mathematically the impedance of this network can be written as:

$$Z(s) = \left(\frac{1}{s\,.\,O_1} + s\,.\,L_1 + R_1\right) \Bigg\| \left(\frac{1}{s\,.\,C_0}\right)$$

or,

$$Z(s) = \frac{s^2 + s\dfrac{R_1}{L_1} + \omega_s^2}{s\left[s^2 + s\dfrac{R_1}{L_1} + \omega_p^2\right]}$$

$$\Rightarrow \omega_s = \frac{1}{\sqrt{L_1.C_1}} \qquad \omega_p = \sqrt{\frac{C_1 + C_0}{L_1\,.\,C_1\,.\,C_0}}$$

where s is the complex frequency ($s = j\omega$), ω_s is the series resonant frequency in radians per second and ω_p is the parallel resonant frequency in radians per second.

Adding additional capacitance across a crystal will cause the parallel resonance to shift downward. This can be used to adjust the frequency that a crystal oscillator oscillates at. Crystal manufacturers normally cut and trim their crystals to have a specified resonant frequency with a known 'load' capacitance added to the crystal. For example, a 6 pF 32 kHz crystal has a parallel resonance frequency of 32,768 Hz when a 6.0 pF capacitor is placed across the crystal. Without this capacitance, the resonance frequency is higher than 32,768 Hz.

Temperature Effects

A crystal's frequency characteristic depends on the shape or 'cut' of the crystal. A tuning fork crystal is usually cut such that its frequency over temperature is a parabolic curve centred around 25 °C. This means that a tuning fork crystal oscillator will resonate close to its target frequency at room temperature, but will slow down when the temperature either increases or decreases from room temperature. A common parabolic coefficient for a 32 kHz tuning fork crystal is "0.04 ppm/°C².

$$f = f_0 \, [1 - 0.04 \text{ ppm } (T - T_0)^2]$$

In a real application, this means that a clock built using a regular 32 kHz tuning fork crystal will keep good time at room temperature, lose 2 minutes per year at 10 degrees Celsius above (or below) room temperature and lose 8 minutes per year at 20 degrees Celsius above (or below) room temperature.

Crystals and Frequency

The crystal oscillator circuit sustains oscillation by taking a voltage signal from the quartz resonator, amplifying it, and feeding it back to the resonator. The rate of expansion and contraction of the quartz is the resonant frequency, and is determined by the cut and size of the crystal.

A regular timing crystal contains two electrically conductive plates, with a slice or tuning fork of quartz crystal sandwiched between them. During startup, the circuit around the crystal applies a random noise AC signal to it, and purely by chance, a tiny fraction of the noise will be at the resonant frequency of the crystal. The crystal will therefore start oscillating in synchrony with that signal. As the oscillator amplifies the signals coming out of the crystal, the crystal's frequency will become stronger, eventually dominating the output of the oscillator. Natural resistance in the circuit and in the quartz crystal filter out all the unwanted frequencies.

One of the most important traits of quartz crystal oscillators is that they can exhibit very low phase noise. In other words, the signal they produce is a pure tone. This makes them particularly useful in telecommunications where stable signals are needed, and in scientific equipment where very precise time references are needed.

The output frequency of a quartz oscillator is either the fundamental resonance or a multiple of the resonance, called an overtone frequency.

A typical Q for a quartz oscillator ranges from 10^4 to 10^6. The maximum Q for a high stability quartz oscillator can be estimated as $Q = 1.6 \times 10^7/f$, where f is the resonance frequency in megahertz.

Environmental changes of temperature, humidity, pressure, and vibration can change the resonant frequency of a quartz crystal, but there are several designs that reduce these environmental effects. These include the TCXO, MCXO, and OCXO. These designs (particularly the OCXO) often produce devices with excellent short-term stability. The limitations in short-term stability are due mainly to noise from electronic components in the oscillator circuits. Long term stability is limited by aging of the crystal.

Due to aging and environmental factors such as temperature and vibration, it is hard to keep even the best quartz oscillators within one part in 10^{-10} of their nominal frequency without constant adjustment. For this reason, atomic oscillators are used for applications that require better long-term stability and accuracy.

Although crystals can be fabricated for any desired resonant frequency, within technological limits, in actual practice today engineers design crystal oscillator circuits around relatively few standard frequencies, such as 3.58MHz, 10 MHz, 14.318, 20 MHz, 33.33 Mhz, and 40 MHz. The vast popularity of the 3.58MHz and 14.318MHz crystals is attributed initially to low cost resulting from scale of economy resulting from the popularity of television and the fact that this frequency is involved in synchronising to the colourburst signal necessary to display colour on an NTSC or PAL based television set. Using frequency dividers, frequency multipliers and phase locked loop circuits, it is possible to synthesise any desired frequency from the reference frequency.

Care must be taken to use only one crystal oscillator source when designing circuits to avoid subtle failure modes of metastability in electronics. If this is not possible, the number of distinct crystal oscillators, PLLs, and their associated clock domains should be rigorously minimised, through techniques such as using a subdivision of an existing clock instead of a new crystal source. Each new distinct crystal source needs to be rigorously justified, since each one introduces new, difficult to debug probabilistic failure modes, due to multiple crystal interactions, into equipment.

Series or Parallel Resonance

A quartz crystal provides both series and parallel resonance. The series resonance is a few kilohertz lower than the parallel one. Crystals below 30 MHz are generally operated at parallel resonance, which means that the crystal impedance appears infinite. Any additional circuit capacitance will thus pull the frequency down. For a parallel resonance crystal to operate at its specified frequency, the electronic circuit has to provide a total parallel capacitance as specified by the crystal manufacturer.

Crystals above 30 MHz (up to >200 MHz) are generally operated at series resonance where the impedance appears at its minimum and equal to the series resistance. For this reason the series resistance is specified (<100 Ω) instead of the parallel capacitance. For the upper frequencies, the crystals are operated at one of its overtones, presented as being a fundamental, 3rd, 5th, or even 7th overtone crystal. The oscillator electronic circuits usually provides additional LC circuits to select the wanted overtone of a crystal.

Spurious Frequencies

For crystals operated in series resonance, significant (and temperature-dependent) spurious responses may be experienced. These responses typically appear some tens of kilohertz above the wanted series resonance. Even if the series resistances at the spurious resonances appear higher than the one at wanted frequency, the oscillator may lock at a spurious frequency (at some temperatures). This is generally avoided by using low impedance oscillator circuits to enhance the series resistance differences.

Notation

On electrical schematic diagrams, crystals are designated with the class letter "Y" (Y1, Y2, etc.) Oscillators, whether they are crystal oscillators or other, are designated with the class letter "G" (G1, G2, etc.)

Crystal oscillator types and their abbreviations:

- ATCXO: Analog Temperature Controlled Crystal Oscillator
- CDXO: Calibrated Dual Crystal Oscillator
- MCXO: Microcomputer-compensated Crystal Oscillator
- OCVCXO: Oven Controlled Voltage Controlled Crystal Oscillator
- OCXO: Oven-Controlled Crystal Oscillator
- RbXO: Rubidium Crystal Oscillators (RbXO), a crystal oscillator (can be a MCXO) synchronised with a built-in rubidium standard which is run only occasionally to save power
- TCVCXO: Temperature Compensated Voltage Controlled Crystal Oscillator
- TCXO: Temperature Compensated Crystal Oscillator
- TSXO: Temperature Sensing Crystal Oscillator, an adaptation of the TCXO
- VCTCXO: Voltage Controlled Temperature Compensated Crystal Oscillator
- VCXO: Voltage-controlled Crystal Oscillator
- DTCXO: Digital Temperature Compensated Crystal Oscillator

Dynatron Oscillator

A dynatron oscillator is a unique electronic circuit that uses negative resistance to keep an LC tank oscillating. In theory, if an ideal capacitor is connected in parallel with an ideal inductor, they form an LC tank circuit that, once it begins oscillating, will oscillate forever as the energy is transferred back and forth between the capacitor and the inductor.

In practice, however, the two components are not ideal. Because "real" components have some associated series resistance (primarily contributed by the inductor), such a circuit is often represented either as an inductor, capacitor and resistor all in series or all in parallel.

If a negative resistance equal in magnitude to this positive resistance can be connected in parallel with the above circuit, then the two resistances will cancel and the circuit will oscillate forever.

Although no passive negative resistors exist, some active devices exhibit a negative resistance characteristic. Examples of such devices are the tetrode thermionic valve (or tube) and the tunnel diode. Using these devices to provide the negative resistance causes the tuned circuit to oscillate as long as power is provided to the active device.

Though the oscillation may appear to be a form of electrical perpetual motion, this is not the case as energy is constantly being provided from the power supply of the active device to sustain the oscillation. In fact, sufficient noise energy is available to allow energy to be coupled out of the oscillating circuit.

This type of oscillator is commonly known as a 'dynatron' oscillator, though, technically, only the version using the tetrode valve is a true dynatron oscillator.

Grid Dip Oscillator

Grid dip oscillator (GDO), also called grid dip metre, dip metre, dipmeter, or just dipper, is a measuring instrument to measure resonant frequency of radio frequency circuits. It measures the amount of absorption of high frequency electromagnetic field by nearby objects. It is an oscillator whose output energy changes in the vicinity of a circuit which is in resonance with the frequency the oscillator generates; somewhat similar to an acoustic tone becoming louder when generated in the vicinity of a resonant cavity or a string tuned to the same frequency.

As an alternative, an absorption wavemeter can be used. The first GDOs were based on vacuum tubes, therefore the grid, as the measured value was the tube's grid current; their use dates back to 1920's. The more modern GDOs are solid-state and more versatile, and are sometimes called gate dip oscillators or emitter dip oscillators.

GDOs were widely used by amateur radio operators for many years, for measuring the properties of resonant circuits, filters, and antennas. A GDO is a tunable LC circuit with a coil that serves as a loose coupling to the measured LC resonant circuit. Resonance is indicated by the dip of the GDO output. The output is usually indicated by a microammeter.

Hartley Oscillator

The Hartley oscillator is an LC electronic oscillator that derives its feedback from a tapped coil in parallel with a capacitor (the *tank circuit*). Although there is no requirement for there to be mutual coupling between the two coil segments, the circuit is usually implemented as such. A Hartley oscillator is essentially any configuration that uses a pair of series-connected coils and a single capacitor. It was invented by Ralph Hartley, who filed for a patent on June 1, 1915 and was awarded patent number 1,356,763 on October 26, 1920.

Hartley oscillators may be series or shunt fed. A Hartley oscillator is made up of the following:

- Two inductors coupled by mutual inductance (may be a two-winding transformer)

- One tuning capacitor

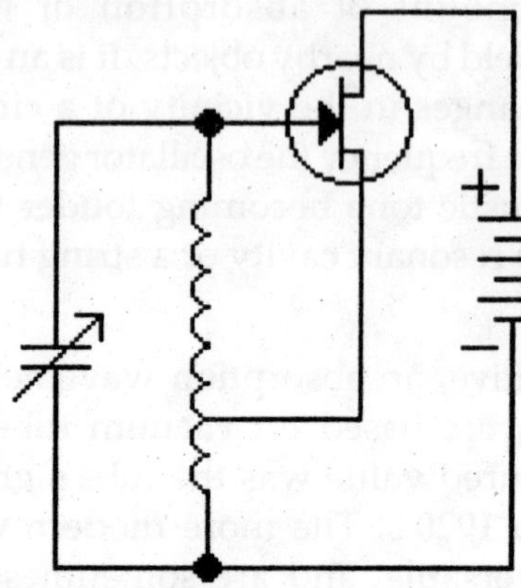

Schematic diagram

Advantages of the Hartley Oscillator include:

- The frequency is varied using a variable capacitor
- The output amplitude remains constant over the frequency range
- The feedback ratio of the tapped inductor remains constant

Disadvantages include:

- Harmonic-rich content of the output
- It is not suitable for a pure sine wave

Uses

The Hartley oscillator was extensively used on all broadcast bands including the FM 88-108MHz band.

Origins

The Hartley oscillator was invented by Ralph V.L. Hartley while he was working for the Research Labratory of the Western Electric Company. Hartley invented the design while overseeing Bell System's transatlantic radiotelephone tests of 1915. In 1946

Hartley was awarded the IRE medal of honour "For his early work on oscillating circuits employing triode tubes and likewise for his early recognition and clear exposition of the fundamental relationship between the total amount of information which may be transmitted over a transmission system of limited bandwidth and the time required." (The second half of the citation refers to Hartley's work in information theory which largely paralelled Harry Nyquist.)

Local Oscillator

A local oscillator is a device used to generate a signal which is beat against the signal of interest to mix it to a different frequency. The oscillator produces a signal which is injected into the mixer along with the signal from the antenna in order to effectively change the antenna signal by heterodyning with it to produce the sum and difference (with the utilisation of trigonometric angle sum and difference identities) of that signal one of which will be at the intermediate frequency which can be handled by the IF amplifier. These are the beat frequencies. Normally the beat frequency is associated with the lower sideband, the difference between the two. Several local oscillators can be strung in series to form a local oscillator chain (LO chain).

Numerically-controlled Oscillator

A numerically controlled oscillator or digitally controlled oscillator (DCO) is an electronic system for synthesising a range of frequencies from a fixed timebase.

Capabilities and Limitations

Unlike a phase-locked loop-based analog frequency synthesiser, it is capable of synthesising a very wide range of precise frequency ratios. However, it is limited to producing frequencies that are less than half the timebase frequency, due to Nyquist's theorem. The name is by analogy with "voltage-controlled oscillator".

Operation

A DCO generally consists of a digital waveform generator that increments a phase counter by a per-sample increment. This phase is then looked up in a waveform table to create a sine waveform. If an analog output is desired, this waveform is sent to a digital-to-analog converter to produce an analog waveform, and then filtered by an analog filter to remove aliasing and DAC artifacts. Alternatively, the digital waveform may be used directly as an input for further digital signal processing.

In this way, the frequency ratios that can be produced are limited only to the precision of the arithmetic used to compute the phase. At the same time, DCOs are phase- and frequency-agile, and can trivially be modified to produce phase-modulated or frequency-modulated outputs, or quadrature outputs.

A DCO can consist of a digital counter such as the 82C53 chip. These produce square wave signals that are very accurate. The square wave can then be modified by standard analog waveshaping techniques. DCOs have been used as cheap replacements for VCOs in some synthesisers. In these applications, their greatest benefit is stability along with having no frequency drift.

Oscillator Sync

Oscillator sync is a feature in synthesisers. One oscillator will restart the period of another oscillator, so that they will have the same base frequency. The timbre can be altered on the synched oscillator by varying its frequency input.

A synched oscillator that resets the other oscillator(s) is called the master, and any synched oscillator that is reset by another oscillator is called a slave.

There are two basic forms of oscillator sync which appear on synthesisers: Hard Sync and Soft Sync

Hard Sync

This form of oscillator sync is more common than soft sync, but frequently generates aliasing. In this form, the master oscillator's pitch is generated by user input (typically the synthesiser's keyboard), and is arbitrary. The slave oscillator's pitch may be tuned to (or detuned from) this frequency, or may remain constant.

Every time the master oscillator's cycle repeats, the slave is retriggered, regardless of its position. If the slave is tuned to a lower frequency than the master it will be forced to repeat before it completes an entire cycle, and if it is tuned to a higher frequency it will be forced to repeat partway through a second or third cycle.

This technique ensures that the oscillators are technically playing at the same frequency, but the irregular cycle of the slave oscillator often causes unnatural timbres, the impression of harmony, and aliasing.

This effect is achieved by measuring the zero axis crossings of the master oscillator and retriggering the slave oscillator after every other crossing.

Soft Sync

This form of oscillator sync is less common, but less frequently generates aliasing. This form is very similar to Hard Sync, with one small difference. In a Hard Sync setup, the slave oscillator is forced to reset to zero with every cycle of the master regardless of position or direction of the slave waveform, which often generates asymmetrical shapes. In Soft Sync, rather than resetting to zero, the wave is inverted; that is, its direction is reversed.

This effect is achieved by measuring the zero axis crossings of the master oscillator and reversing the slope of the slave oscillator after every other crossing.

Due to the rarity and ill-defined nature of Soft Sync, several other effects may also be referred to by the same:

- Hard Sync which is disabled when the frequency or amplitude of the slave crosses a user-defined threshold.
- Hard Sync which is disabled when the frequency of the slave extends too high above or too far below the frequency of the master.
- Hard Sync which is disabled when the frequency of the slave is lower than the frequency of the master.

Soft Sync may accurately refer to any of these, depending on the synthesiser or manufacturer in question. However, because none of them actually synthesise the waveform in a way different from Hard Sync (rather, they selectively deactivate it) they are listed here as secondary.

Wien Bridge Oscillator

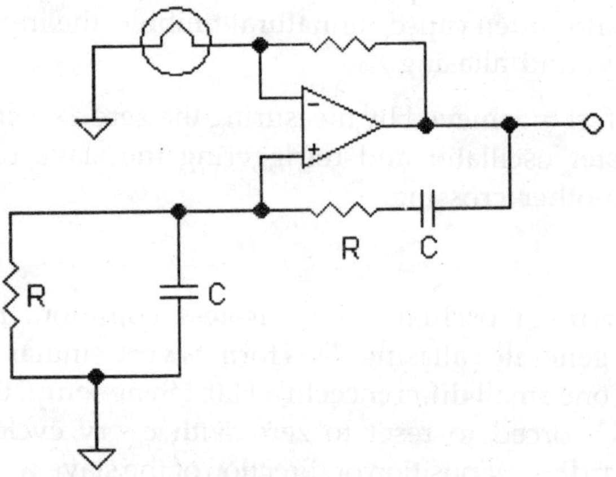

Classic Wien bridge oscillator

A Wien bridge oscillator is a type of electronic oscillator that generates sine waves without having any input source. It can output a large range of frequencies. The bridge comprises four resistors and two capacitors. The circuit is based on a network originally developed by Max Wien in 1891. At that time, Wien did not have a means of developing electronic gain so a workable oscillator could not be realised. The modern circuit is derived from William Hewlett's 1939 Stanford University master's degree thesis. Hewlett, along with David Packard co-founded Hewlett-Packard. Their first product was the HP 200A, a precision sine wave oscillator based on the Wien bridge. The 200A is a classic instrument known for its low distortion.

The frequency of oscillation is given by:

$$f = \frac{1}{2\pi RC}$$

Amplitude Stabilisation

The key to Hewletts' low distortion oscillator is effective amplitude stabilisation. The amplitude of electronic oscillators tends to increase until clipping or other gain limitation is reached. This leads to high harmonic distortion, which is often undesirable.

Hewlett used an incandescent bulb in the oscillator feedback path to limit the gain. The resistance of light bulbs and similar heating elements increases as their temperature increases. If the oscillation frequency is significantly higher than the thermal time constant of the heating element, the radiated power is proportional to the oscillator power. Since heating elements are black body radiators, they follow the Stefan-Boltzmann law. The radiated power is proportional to T^4, so resistance increases at a greater rate than amplitude. If the gain is inversely proportional to the oscillation amplitude, the oscillator gain stage reaches a steady state and operates as a near ideal class A amplifier, achieving very low distortion.

Modern Wien bridge oscillators have used field effect transistors or photocells for amplitude stabilisation in place of light bulbs. Distortion as low as 0.0008% (8 parts per million) can be achieved with only modest improvements to Hewlett's original circuit.

Another way of stabilising the amplitude is by using non-linear elements, such as diodes, to modify the resistance of the negative feedback network. An example of use can be seen at (in Spanish).

Analysis

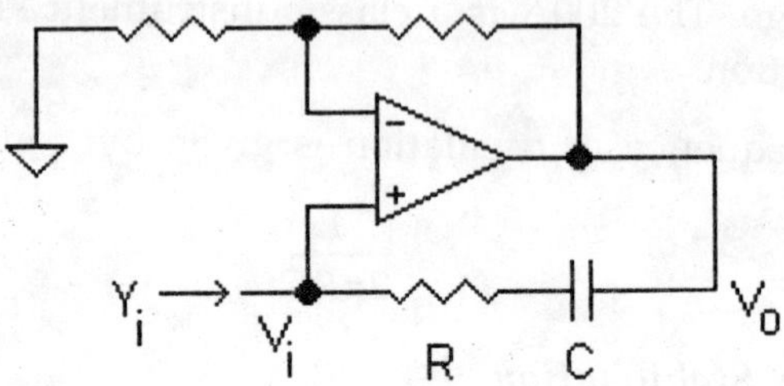

Input admittance analysis

If a voltage source is applied directly to the input of an ideal amplifier with feedback, the input current will be:

$$i_{in} = \frac{v_{in} - v_{out}}{Z_f}$$

Where v_{in} is the input voltage, v_{out} is the output voltage, and Z_f is the feedback impedance. If the voltage gain of the amplifier is defined as:

$$A_v = \frac{v_{out}}{v_{in}}$$

And the input admittance is defined as:

$$Y_i = \frac{i_{in}}{v_{in}}$$

Input admittance can be rewritten as:

$$Y_i = \frac{1 - A_v}{Z_f}$$

For the Wien bridge, Z_f is given by:

$$Z_f = R + \frac{1}{j\omega C}$$

Substituting and rewriting:

$$Y_i = \frac{(1-A_v)\left(\omega^2C^2R + j\omega C\right)}{1+(\omega CR)^2}$$

If A_v is greater than 1, the input admittance is a negative resistance in parallel with an inductance. The inductance is:

$$L_{in} = \frac{\omega^2C^2R^2+1}{\omega^2C^2(A_v-1)}$$

If a capacitor with the same value of C is placed in parallel with the input, the circuit has a natural resonance at:

$$\omega = \frac{1}{\sqrt{L_{in}C}}$$

Substituting and solving for inductance yields:

$$L_{in} = \frac{R^2C}{A_v-2}$$

If A_v is chosen to be 3:

$$L_{in} = R^2C$$

Substituting this value yields:

$$\omega = \frac{1}{RC}$$

Or

$$f - \frac{1}{2\pi RC}$$

Similarly, the input resistance at the frequency above is:

$$R_{in} = \frac{-2R}{A_v-1}$$

For $A_v = 3$:

$R_{in} = -R$

If a resistor is placed in parallel with the amplifier input, it will cancel some of the negative resistance. If the net resistance is negative, amplitude will grow until clipping occurs. Similarly, if the net resistance is positive, oscillation amplitude will decay. If a resistance is added in parallel with exactly the value of *R*, the net resistance will be infinite and the circuit can sustain stable oscillation at any amplitude allowed by the amplifier.

Notice that increasing the gain makes the net resistance more negative, which increases amplitude. If gain is reduced to exactly 3 when a suitable amplitude is reached, stable, low distortion oscillations will result. Amplitude stabilisation circuits typically increase gain until a suitable output amplitude is reached. As long as *R*, *C*, and the amplifier are linear, distortion will be minimal.

Variable-frequency Oscillator

A variable frequency oscillator (VFO) is a component in a radio receiver or transmitter that controls the frequency to which the apparatus is tuned. It is a necessary component in any radio receiver or transmitter that works by the superheterodyne principle, and which can be tuned across various frequencies.

Why do radios need a VFO?

In a simple superhet radio receiver, incoming radio frequencies from the antenna are made to mix (or multiply) with an internally generated radio frequency from the VFO in a process called mixing.

The mixing process can produce a range of output signals:

- At all the original frequencies,
- At frequencies that are the sum of each two mixed frequencies
- At frequencies that equal the difference between two of the mixed frequencies
- At other, usually higher, frequencies.

If the required incoming radio frequency and the VFO frequency were both rather high (RF) but quite similar, then by far the lowest frequency produced from the mixer will be their difference. In very simple radios, it is relatively straightforward to separate this from all the other spurious signals using a filter, to amplify it and then further to process it into an audible signal. In more complex situations, many enhancements and complications get added to this simple process, but this mixing or heterodyning principle remains at the heart of it.

There are two main types of VFO in use: analog and digital.

Analog VFO

An analog VFO could be an electronic oscillator where the value of at least one of the active components is adjustable under user control so as to alter its output frequency. The active component whose value is adjustable is usually a capacitor, but could be a variable inductor.

Tuning Capacitor

The variable capacitor is a mechanical device in which the separation of a series of interleaved metal plates is physically altered to vary its capacitance. Adjustment of this capacitor is sometimes facilitated by a mechanical step-down gearbox to achieve fine tuning.

Varactor

A reversed-biased semiconductor diode exhibits capacitance. Since the width of its non-conducting depletion region depends on the magnitude of the reverse bias voltage, this voltage can be used to control the junction capacitance. The varactor bias voltage may be generated in a number of ways and there may need to be no significant moving parts in the final design. Varactors have a number of disadvantages including temperature drift and aging, electronic noise, low Q factor and non-linearity.

Digital VFO

Modern radio receivers and transmitters usually use some form of digital frequency synthesis to generate their VFO signal. The advantages of this are manifold, including smaller designs, lack of moving parts, and the ease with which preset frequencies can be stored and manipulated in the digital computer that is usually embedded in the design for other purposes.

It is also possible for the radio to become extremely frequency-agile in that the control computer could alter the radio's tuned frequency many tens, thousands or even millions of times a second.

This capability allows communications receivers effectively to monitor many channels at once, perhaps using digital selective calling (DSC) techniques to decide when to open an audio output channel and alert users to incoming communications. Pre-programmed frequency agility also forms the basis of some military radio encryption and stealth techniques. Extreme frequency agility lies at the heart of spread spectrum techniques that are currently gaining mainstream acceptance in computer wireless networking such as Wi-Fi.

There are disadvantages to digital synthesis such as the inability of a digital synthesiser to tune smoothly through all frequencies, but with the channelisation of many radio bands, this can also be seen as an advantage in that it prevents radios from operating in between two recognised channels.

Digital frequency synthesis almost always relies on crystal controlled frequency sources. Crystal controlled oscillators have enormous advantages over inductive and capacitively controlled ones in terms of stability and repeatability as well as low noise and high Q factor. The disadvantage comes when you try to alter the resonant frequency to tune the radio, but a wide range of digital techniques have made this unnecessary in modern practice.

Digital Frequency Synthesis

The electronic and digital techniques involved in this include:

- *Direct Digital Synthesis (DDS):* Enough data points for a mathematical sine function are stored in digital memory. These are recalled at the right speed and fed to a digital to analog converter where the required sine wave is built up.
- *Direct Frequency Synthesis:* Early channelised communication radios had multiple crystals - one for each channel on which they could operate. After a while this thinking was combined with the basic ideas of heterodyning and mixing described under Why do radios need a VFO? Multiple crystals can be mixed in various combinations to produce various output frequencies.
- *Phase Locked Loop (PLL):* Using a varactor-controlled or voltage-controlled oscillator (VCO) (described above in varactor under analog VFO techniques) and a phase detector, a control-loop can be set up so that the VCO's output is frequency-locked to a crystal controlled reference oscillator. This would not be much use unless the phase detector's comparison were made not between the actual outputs of the two oscillators, but between the outputs of each after frequency division by two slightly different divisors. Then by altering the frequency-division divisor(s) under computer control, a variety of actual (undivided) VCO output frequencies can be generated.

It is this last, the PLL technique, that dominates most radio VFO design thinking today.

Performance

The performance of a radio's VFO strongly influences the performance of the radio itself.

Accuracy

It is useful if the frequency produced by the VFO is both stable and repeatable.

Stability

An unstable VFO's output frequency will drift with time. The root cause of this can often be traced to temperature dependency in some of the voltages and component values involved. Often as radios warm up it is necessary to slightly retune them to remain on frequency.

Repeatability

Ideally, for the same selected radio channel, the VFO should generate exactly the same frequency today as it was on the day the radio was first assembled and tested. This will mean that any built-in errors seen that day during the manufacture will have been calibrated out, and this calibration will not have changed. If this is not the case, then you will not be able entirely to trust your tuning dial.

An erratic VFO would be a source of irritation on a receiver, as one would have to tune slightly off the known frequency to receive a certain station. The problem can be more serious in a transmitter as the operator could be held responsible for unwittingly and illegally transmitting on a frequency for which he or she is not authorised or licensed.

Purity

You can imagine the shape of the VFO's frequency vs. amplitude graph to be the shape of the 'window' through which the radio receives (and in the case of a transmitter, through which it transmits when you ask it to transmit a pure sine-wave tone). In the ideal case, this frequency/amplitude plot is very simple, i.e. there is absolutely no output at any frequency except one, and plenty of pure output at exactly that frequency. In this ideal case, of course, the 'window' is unique and infinitely narrow. The ideal radio will receive and transmit only exactly what is expected.

Spurii

A VFO's frequency vs. amplitude graph (or Fourier Analysis) may exhibit not one but several narrow peaks, probably harmonically related. Each of these other peaks can potentially mix with some other incoming signal and produce a *spurious* response. These *spurii* (sometimes spelt *spuriae*) result in you hearing two stations at once, even though the other is nowhere near this one on the band.

The extra peaks may be many hundreds or thousands of times lower in value than the main one, but don't forget that the other, interfering station may be hundreds or thousands of times more powerful at the antenna than the one you are after.

In a transmitter, these spurious signals are actually generated along with the one you expect. If they are not completely filtered out before they are transmitted, then the license-holder may again be in breach of the terms of his or her license.

Phase Noise

When examined with very sensitive equipment, the pure sine-wave peak in a VFO's frequency graph will most likely turn out not to be sitting on a flat noise-floor. Slight random 'jitters' in the signal's timing will mean that the peak is sitting on 'skirts' of phase-noise at frequencies either side of the desired one.

These are also troublesome in crowded bands. They allow through unwanted signals that are fairly close to the one we expect, but because of the random quality of these phase-noise 'skirts', the signals are usually unintelligible, appearing just as extra noise in the signal we are after. The effect is that what should be a clean signal in a crowded band can appear to be a very noisy signal, because of the effects of all the strong signals nearby.

The effect of VFO phase noise on a transmitter is that

random noise is actually transmitted either side of the required signal. Again, this must be avoided at all costs for legal reasons in many cases.

Crystal Control

In all performances cases, crystal controlled oscillators are better behaved than the semiconductor- and LC-based alternatives. They tend to be more stable, more repeatable, have fewer and lower harmonics and lower noise than all the alternatives in their cost-band. This in part explains their huge popularity in low-cost and computer-controlled (i.e. PLL and synthesiser-based) VFOs.

Relaxation Oscillator

A relaxation oscillator is an oscillator in which a capacitor is charged gradually and then discharged rapidly. It's usually implemented with a resistor or current source, a capacitor, and some sort of "threshold" device such as a, neon lamp, diac, unijunction transistor, or Gunn diode. For simplification below, a single "threshold" device will be replaced by a set of comparators and a SR Latch.

The capacitor is charged through the resistor, causing the voltage across the capacitor to approach the charging voltage on an exponential curve. In parallel with the capacitor is the threshold device. Such devices don't conduct at all until the voltage across them reaches some threshold (trigger) voltage. They then conduct heavily, quickly discharging the capacitor. When the voltage across the capacitor drops to some lower threshold voltage, the device stops conducting and the capacitor can begin charging again, repeating the cycle. If the threshold element is a neon lamp, the circuit also provides a flash of light with each discharge of the capacitor.

The electrical output of a relaxation oscillator is usually a sawtooth wave. If only a small portion of the exponential ramp is used (that is, if the triggering voltage of the threshold device is much lower than the charging voltage source), the ramp will approximate a linear ramp but if a truly linear sawtooth is required, then the charging resistor should be replaced by some sort of constant current source.

Phase-shift Oscillator

A phase shift oscillator is a simple sine wave electronic oscillator. It contains an inverting amplifier, and a feedback filter which 'shifts' the phase by 180 degrees at the oscillation frequency.

The filter must be designed so that at frequencies above and below the oscillation frequency, the signal is shifted by either more or less than 180 degrees. This results in constructive superposition for signals at the oscillation frequncies, and destructive superposition for all other frequencies.

The most common way of achieving this kind of filter is using 3 cascaded resistor-capacitor filters, which produce no phase shift at one end of the frequency scale, and a phase shift of 270 degrees at the other end.

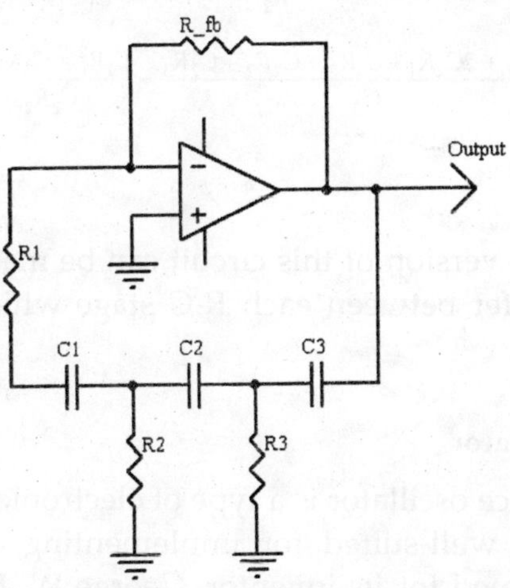

One of the simplest implementations for this type of oscillator uses an operational amplifier (op-amp), 3 capacitors and 4 resistors, as shown in the diagram.

The mathematics for calculating the oscillation frequency and oscillation criteria for this circuit are surprisingly complex, due to each R-C stage loading the previous ones. The calculations are greatly simplified by setting all the resistors (except the negative feedback resistor) and all the capacitors to the same values. In the diagram, if R1 = R2 = R3 = R, and C1 = C2 = C3 = C, then:

$$f_{Oscillation} = \frac{1}{2\pi RC\sqrt{6}}$$

and

Oscillation criteria: $R_{feedback} = 29R$

Without the simplification of all the resistors and capacitors having the same value, the calculations become more complex:

$$f_{Oscillation} = \frac{1}{2\pi\sqrt{R_2R_3(C_1C_2 + C_1C_3 + C_2C_3) + R_1R_3(C_1C_2 + C_1C_3) + R_1R_2C_1C_2}}$$

Oscillation criteria:

$$f_{Oscillation} = 2(R_1 + R_2 + R_3) + \frac{2R_2R_3}{R_2} + \frac{C_2R_2 + C_2R_3 + C_3R_3}{C_1} + \frac{C_1R_1 + C_1R_2 + C_3R_3}{C_2}$$

$$+ \frac{2C_1R_1 + 2C_2R_1 + C_1R_2 + C_2R_2 + C_2R_3}{C_3} + \frac{C_1R_1^2 + C_3R_1R_3}{C_2R_2} + \frac{C_2R_1R_3 + C_1R_1^2}{C_3R_2} +$$

$$\frac{C_1R_1^2 + C_1R_1R_2 + C_2R_1R_2}{C_3R_3}$$

Another version of this circuit can be made by putting an op-amp buffer between each R-C stage which simplifies the calculations.

Pierce Oscillator

The Pierce oscillator is a type of electronic oscillator circuit particularly well-suited for implementing crystal oscillator circuits. Named for its inventor, George W. Pierce, the Pierce oscillator is a derivative of the Colpitts oscillator. Virtually all digital IC clock oscillators are of Pierce type, as the circuit can be implemented using a minimum of components: a single digital inverter, two resistors, two capacitors, and the quartz crystal, which acts as a highly selective filter element. The low manufacturing cost of this circuit, combined with the outstanding frequency stability of the quartz crystal, give it an advantage over other designs in many consumer electronics applications.

Simple Pierce Oscillator

R_1 acts as a feedback resistor, causing the inverter to function as an amplifier. The crystal in combination with C_1 and C_2 forms a pi network band-pass filter, which provides a 180 degree phase shift from the output to input at approximately the resonant frequency of the crystal. To understand the operation of this, it can be noted that at the frequency of oscillation, the crystal appears inductive; thus it can be considered a large inductor with a high Q. The combination of the phase shift from the pi network and the negative gain from the inverter results in a positive loop gain and thus positive feedback, leading to oscillation. A second resistor could be used between the output of the inverter and the crystal to isolate the inverter from the crystal network. This would also add additional phase shift to C_1.

Load Capacitance

The total capacitance seen from the crystal looking into the rest of the circuit is called the "load capacitance". When a manufacturer makes a "parallel" crystal, a technician uses a Pierce oscillator with a particular load capacitance (often 18 or 20 pF) while trimming the crystal to oscillate at exactly the frequency written on its package.

To get the same frequency performance, an engineer must then make sure that the capacitances in the circuit match this value specified in the crystal's data sheet. Load capacitance C_L can be calculated from the series combination of C_1 and C_2 (including stray capacitances from the oscillator, PCB layout, and crystal case), in parallel with any stray shunt capacitance between the two leads C_S (often assumed to be 3-9 pF):

$$C_L = (C_1 inserieswithC_2) \parallel C_s = \frac{C_1 \times C_2}{C_1 - C_2} + C_s = \frac{1}{\frac{1}{C_1} + \frac{1}{C_2}} - C_s$$

When a manufacturer makes a "series" crystal, a technician uses a different tuning procedure. When such a crystal is used in a Pierce oscillator, the Pierce oscillator (as always) drives the crystal at nearly its parallel resonance frequency. But that frequency is few kilohertz higher than the series resonant frequency printed on the package of a "series" crystal.

Increasing the "load capacitance" slightly decreases the frequency generated by a Pierce oscillator, but never enough to reduce it all the way down to the series resonant frequency.

6

Electrical Oscillations

Circuit Transients

Transient Phenomenon: When a circuit is switched from one condition to another by a change in applied voltage, there is a transitional period before the circuit reaches a steady state. The duration of this transitional period depends upon the parameters of the system. The behaviour of the system during the transitional period is known as the 'transient effect' and the phenomenons is known as 'transient phenomenon'.

Let us consider the transient phenomenon in an electrical circuit when a DC voltage is impressed on the circuit, and also when the DC voltage is withdrawn.

Growth of Current in an L-R circuit when DC voltage is Switched on: Let us consider a circuit containing an inductance *L*, a resistance *R* and a cell of *constant* e.m.f. *E*. When the circuit is closed by throwing the switch *S* to *a*, a self-induced e.m.f. is set up in *L* which opposes the growth of current in the circuit. Hence the current does not reach its final steady value *E/R* instantaneously, but grows at a rate depending upon the inductance and resistance of the circuit.

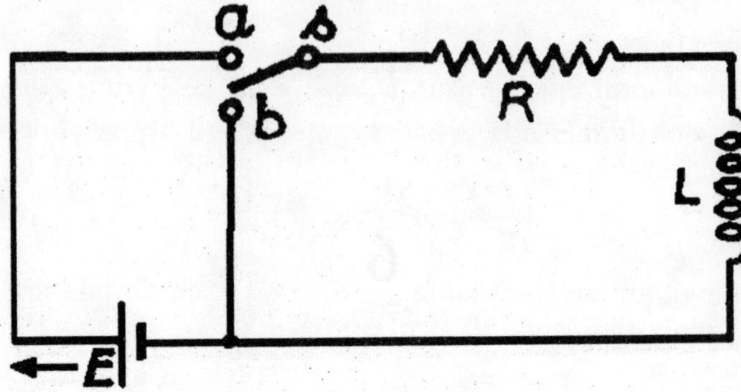

During the transient state when the current is growing, let i be the current at any instant t. The opposing induced e.m.f. will be E – L(*di/dt*). Hence the effective e.m.f. driving the current in the circuit at the instant will be *E-L (di/dt)*. This, by Ohm's law, must be equal to *Ri*. That is

$$E - L\frac{di}{dt} = Ri$$

or
$$E - Ri = L\frac{di}{dt}$$

or
$$dt = \frac{L\,di}{E - Ri}$$

Integrating, we get

$$t = -\frac{L}{R}\log_e (E - Ri) + C,$$

where C is the integration constant.

$$\text{At } t=0,\ i=0, \quad \therefore \quad C = \frac{L}{R}\log_e E$$

$$t = -\frac{L}{R}\log_e (E - Ri) + C, + \frac{L}{R}\log_e E$$

or
$$-\frac{L}{R}t = \log_e (E - Ri) - \log_e E = \log_e \frac{E - Ri}{E}$$

or $$e^{-(R/L)t} = \frac{E - Ri}{E} = 1 - \frac{Ri}{E}$$

or $$\frac{Ri}{E} = [1 - e^{-(R/L)t}]$$

or $$i = \frac{Ri}{E}[1 - e^{-(R/L)t}]$$

or $$\boxed{i = i_0\left[1 - e^{-(R/L)t}\right]} \quad ...(i)$$

where i_o ($=E/R$) is the final steady value of the current reached when $t \to \infty$. Equation (i) shows that the current in the circuit rises according to an exponential law.

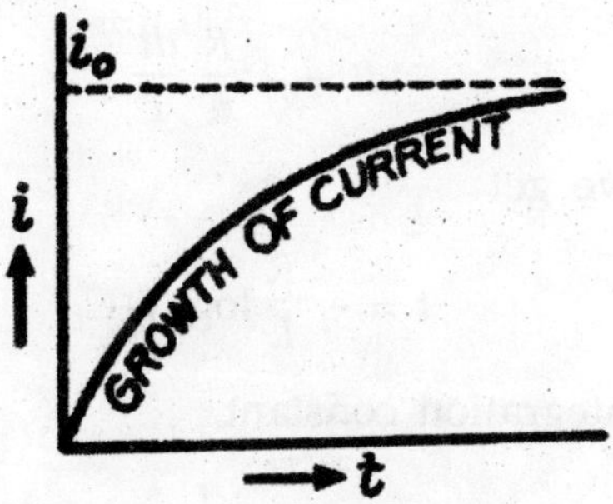

The rate of growth of the current is

$$\frac{di}{dt} = i_0 \frac{L}{R} e^{-(R/L)t}$$

$$= i_0 \frac{L}{R}\left(1 - \frac{i}{i_0}\right) \text{ [from eq. (i)]}$$

$$= \frac{L}{R}(i - i_0).$$

Thus it is seen that greater the ratio *R/L*, or smaller the ratio. *L/R*, the more rapidly does the current approach its maximum value. The ratio *L/R* is called the 'inductive time-constant' of the circuit and is expressed in second if *L* and *R* are given in henry and ohm respectively.

Decay of Current when DC Voltage is Switched off

When the switch S is thrown over to *b*, the e.m.f. *E* applied to the circuit becomes zero, yet the resistance of the circuit remains unchanged. The self-induction of the coil cow opposes the fall of the current. During the variable state when the current is falling, let i be the current at any instant *t*. Since now there is no applied e.m.f. in the circuit, the p.d. *Ri* across the resistance *R* is equal and opposite to the induced e.m.f. *L* (di/dt). Hence

$$-L\frac{di}{dt} = Ri$$

or $$dt = -\frac{R}{L}\frac{di}{i}.$$

Integrating, we get

$$t = -\frac{R}{L}\log_e i + C,$$

where *C* is integration constant.

$$\text{At } t=0,\ i=i_0, \quad \therefore\ C = \frac{L}{R}\log_e E. \text{ Hence}$$

$$t = -\frac{R}{L}\log_e i + \frac{R}{L}\log_e i_0$$

or $$-\frac{R}{L}t = \log_e i - \log_e i_0 = \log_e \frac{i}{i_0}$$

or $$e^{-(R/L)t} = i/i_0$$

or $$\boxed{i = i_0 e^{-(R/L)t}} \qquad \text{...(ii)}$$

Equation (ii) shows that the current in the circuit decays exponentially. The rate of fall of current is

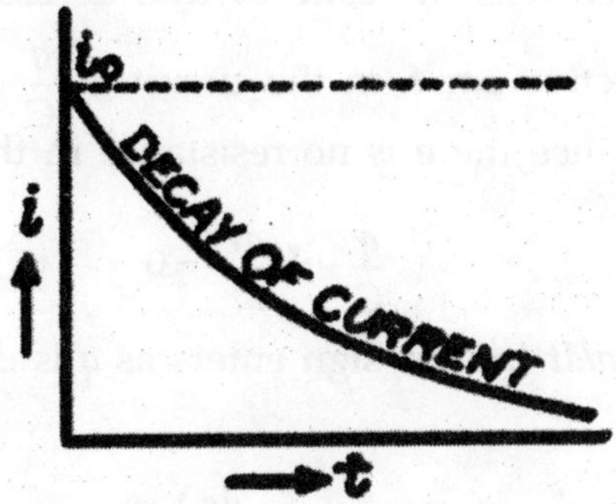

$$\frac{di}{dt} = -i_0 \frac{R}{L} e^{-(R/L)t}$$

$$= -i_0 \frac{R}{L} \frac{i}{i_0} \quad \text{[From eq. (ii)]}$$

$$= -\frac{R}{L} i.$$

Thus it is clear that greater the ratio *R/L*, or smaller the time-constant *L/R*, the more rapidly does the current die away.

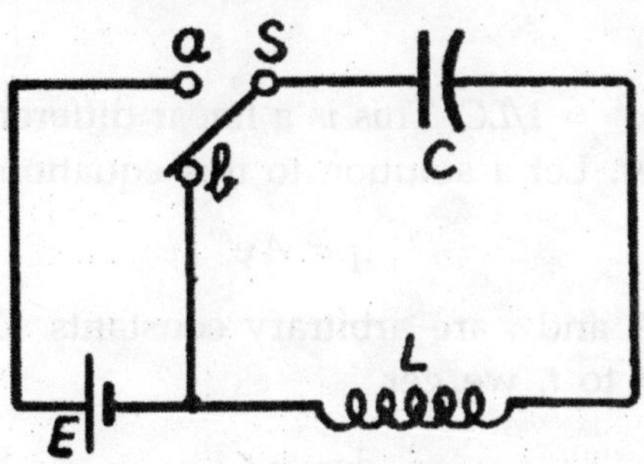

Discharge of a Capacitor through Inductance: Oscillatory L-C Circuit (Electrical Oscillations)

Let a capacitor of capacitance *C*, carrying a charge q_0, be discharged through a coil of inductance *L* and negligible resistance. At an instant *t* during the discharge, let *q* be the charge left on the capacitor and *i* the discharging current through the inductance. Then the potential difference across the capacitor is *q/C* and the induced e.m.f. in the inductance is $L(di/dt)$ which acts opposite to that across the capacitor. Hence the effective e.m.f. in the circuit is $\frac{q}{C} - L\frac{di}{dt}$. But this must be zero since there is no resistance in the circuit. Thus

$$\frac{q}{C} - L\frac{di}{dt} = 0$$

But $i = -dq/dt$ (minus sign enters as q is decreasing with time).

$$\therefore \quad \frac{q}{C} - L\frac{d}{dt}\left(-\frac{dq}{dt}\right) = 0$$

or
$$L\frac{d^2q}{dt^2} + \frac{q}{C} = 0$$

or
$$\frac{d^2q}{dt^2} + \frac{q}{LC} = 0,$$

or
$$\boxed{\frac{d^2q}{dt^2} + \omega^2 q = 0} \quad ..(i)$$

where $\omega^2 = 1/LC$. This is a linear differential equation of second order. Let a solution to this equation be

$$q = Ae^{\alpha t},$$

where A and α are arbitrary constants. Differentiating it with respect to t, we get

$$\frac{dq}{dt} = Ae^{\alpha t},$$

and
$$\frac{d^2q}{dt^2} = A\alpha^2 e^{\alpha t}.$$

Substituting for $\frac{d^2q}{dt^2}$ and q in eq. (i). we get

$$A\alpha^2 e^{\alpha t} + \omega^2 A e^{\alpha t} = 0$$

or $$A e^{\alpha t}\left(\alpha^2 + \omega^2\right) = 0$$

or $$\alpha^2 + \omega^2 = 0$$

or $$\alpha = \pm j\omega$$

where $j = \sqrt{(-1)}$. Thus the eq. (i) has two solutions which are

$$q = Ae^{+j\omega t} \text{ and } q = Ae^{-j\omega t}.$$

As the equation is linear, any linear combination of these two solutions will also satisfy it. Hence the most general solution is

$$q = A_1 e^{+j\omega t} + q = A_2 e^{-j\omega t}, \qquad \text{...(ii)}$$

where A_1 and A_2 are new arbitrary constants.

At $t = 0$; $q = q_0$, so that eq. (ii) gives

$$q_0 = A_1 + A_2. \qquad \text{...(iii)}$$

Differentiating eq. (ii), we get

$$i = \frac{dq}{dt} = A_1 e^{j\omega t}(j\omega) + A_2 e^{-j\omega t}(-j\omega).$$

At $t=0$, $i=0$, and so we get

$$0 = A_1 j\omega - A_2 j\omega$$

$$\therefore \quad A_1 = A_2.$$

Therefore eq. (iii) gives

$$A_1 = A_2 = \frac{q_0}{2}.$$

Substituting in eq. (ii), we get

$$q = \frac{q_0}{2}\left(e^{j\omega t} + e^{-j\omega t}\right)$$

or $$\boxed{q = q_0 \cos \omega t.} \qquad \text{...(iv)}$$

This is the equation of the discharge of the capacitor. It shows that the discharge of the capacitor is oscillatory and simple harmonic. The period of oscillation is given by

$$T = \frac{2\pi}{\omega} = 2\pi\sqrt{(LC)},$$

and the frequency of oscillation is

$$n = \frac{1}{T} = \frac{1}{2\pi}\sqrt{\left[\frac{1}{LC}\right]}$$

Energy Change in Electrical Oscillations: Eq. (iv) shows that when a capacitor is discharged through an inductance, electrical oscillations of constant amplitude q_0 (the initial charge on the capacitor) are produced in the circuit. The production of oscillations can be explained in the following way :

In the beginning when the capacitor is fully charged, an amount of energy $\frac{1}{2}\frac{q_0^2}{C}$ is stored in the electric field between its plates. When the circuit is closed, the capacitor begins to discharge through the inductance, causing a current to flow in the anticlockwise direction.

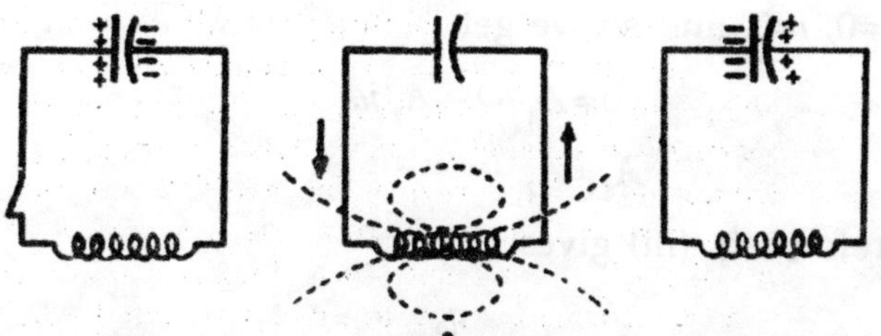

(a) (b) (c)

As the current rises from zero it builds up a magnetic field around the inductance, and when it has reached its maximum value i_0, the energy stored in the magnetic field is $\frac{1}{2} L i_0^2$. At the same time, the capacitor has completely discharged and the p.d. between its plates has reduced to zero. Thus the energy of the electric field between the capacitor plates has transferred to the magnetic field. The current, however, does not stop instantaneously but continues to flow for some time while the magnetic field dies down. By Lenz's law, the dying magnetic field induces an e.m.f. in the inductance *in the same direction as the current*. The current therefore persists, and now charge the capacitor in the opposite sense until the magnetic field has disappeared. Thus the energy is once again stored as $\frac{1}{2}\frac{q_0^2}{C}$ in the electric field between the capacitor plates. The process now repeats itself in the reverse direction and the charge on the capacitor surges back and forth indefinitely. In other words, electrical oscillations take place indefinitely.

This is, however, an ideal case when the inductance is supposed to have zero resistance. In practice, the inductance always has a finite resistance which causes gradual dissipation of the electrical energy as heat. Owing to this damping effect the oscillations gradually diminish in amplitude and ultimately die away.

Analogy with Linear Harmonic Oscillator: The *L–C* oscillations are similar to the oscillations of a mass-spring system. In each case the energy alternates between two form: magneto and electric for the *L-C* system, and kinetic and potential for the mass-spring system. The current in the *L-C* circuit is zero when the entire energy is stored as electric energy in the capacitor, and is maximum when the entire energy is stored as magnetic in the indictor. In the same way, the velocity of mass is zero when the energy is entirely potential energy in the spring, and is maximum when the energy is entirely kinetic energy of the mass. Thus the capacitor is like a spring $(C \equiv 1/k)$ and the indictor is like a mass $(L \equiv m)$. The charge corresponds to the displacement $(q \equiv x)$ and the current corresponds to the velocity $(i \equiv v)$. Further

$$\tfrac{1}{2}Li^2 \equiv \tfrac{1}{2}mv^2$$

and $$\frac{1}{2}\frac{q^2}{C} \equiv \frac{1}{2}kx^2.$$

Use of L-C Circuit: As the capacitor discharges, the energy of the charged capacitor is stored in the magnetic field of the inductance, then again back in the electric field between the capacitor plates, and soon. If the fields are made to alternate rapidly, some energy escapes from the circuit in the form of electromagnetic waves. These waves form the basis of wireless telegraphy.

Oscillatory L-C-R Circuit (Damped Electrical Oscillations)

A circuit containing an inductance, a capacitance and a resistance in series is an example of a damped harmonic oscillator.

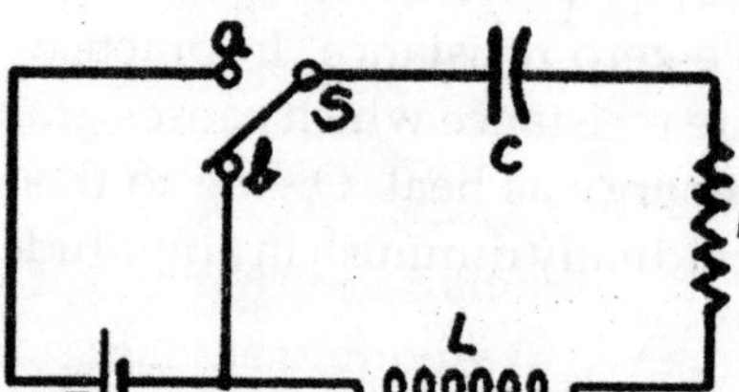

Figure above shows a circuit containing a capacitance *C*, an inductance *L* and a resistance *R* joined in series. *E* is a cell and *S* a switch connected as shown. When *S* is thrown over to *a*, the capacitor acquires a charge say, q_o.

If now the switch is thrown over to *b*, the capacitor discharges through the inductance L and the resistance *R* so that a varying current flows through them. At an instant *t* during the process of discharge, let *q* be the charge left on the capacitor and *i* the current. The induced e.m.f set up in the inductance is L *(di/dt)* acting *opposite* to the p.d. across the capacitor plates which is *q/C*. Thus the effective e.m. f. in the circuit is $\frac{q}{C} - L\frac{di}{dt}$ which is driving the current through the resistance *R*. Hence it must be equal to *Ri*. Thus

$$\frac{q}{C} - L\frac{di}{dt} = Ri.$$

But $i = -dq/dt$ (minus sign enters since q is decreasing with time).

$$\therefore \quad \frac{q}{C} - L\frac{d}{dt}\left(-\frac{dq}{dt}\right) = R\left(-\frac{dq}{dt}\right)$$

or $$L\frac{d^2q}{dt^2} + R\frac{dq}{dt} + \frac{q}{C} = 0$$

or $$\frac{d^2q}{dt^2} + \frac{R}{L} + \frac{dq}{dt} + \frac{q}{LC} = 0.$$

Putting $\frac{R}{L} = 2r$ and $\frac{1}{LC} = \omega^2$, we get

$$\boxed{\frac{d^2q}{dt^2} + 2r\frac{dq}{dt} + \omega^2 q = 0.} \qquad \text{...(i)}$$

This is exactly similar to the equation of damped harmonic oscillator. Here we have the charge q on the capacitor, instead of displacement x of a particle. Let a solution to this equation be

$$q = Ae^{\alpha t}$$

so that $$\frac{dq}{dt} = A\alpha e^{\alpha t}$$

and $$\frac{d^2q}{dt^2} = A\alpha^2 e^{\alpha t}.$$

Substituting in eq. (i), we get

$$A\alpha^2 e^{\alpha t} + 2r\alpha\, Ae^{\alpha t} + \omega^2\, Ae^{\alpha t} = 0$$

or $$\alpha^2 + 2r\alpha + \omega^2 = 0$$

or $$= -r \pm \sqrt{(r^2 - \omega^2)}\,t,$$

Hence the most general solution of eq. (i) is

$$q = A_1 e^{\left[-r+\sqrt{(r^2-\omega^2)}t\right]} + A_2 e^{\left[-r-\sqrt{(r^2-\omega^2)}t\right]}. \quad ...(ii)$$

where A_1 and A_2 are arbitrary constants to be determined by the initial conditions.

Now at $t=0$, $q=q_0$. Therefore from eq. (ii), we have

$$q_o = A_1 + A_2. \quad ...(iii)$$

Differentiating eq. (ii), we get

$$i = \frac{dq}{dt} = A_1\left[-r+\sqrt{(r^2-\omega^2)}\right]e^{\left[-r-\sqrt{(r^2-\omega^2)}\right]t}$$

$$+ A_2\left[-r-\sqrt{(r^2-\omega^2)}\right]e^{\left[-r-\sqrt{(r^2-\omega^2)}\right]t}$$

Applying the condition $i=0$ at $t—0$, we get

$$0 = A_1\left[-r+\sqrt{(r^2-\omega^2)}\right] + A_2\left[-r-\sqrt{(r^2-\omega^2)}\right]$$

$$= -r(A_1+A_2) + \sqrt{(r^2-\omega^2)}\,+(A_1-A_2)$$

$$= rq_0 + \sqrt{(r^2-\omega^2)}\,(A_1-A_2) \quad \text{[from eq. (iii)]}$$

$$\therefore\ A_1 - A_2 = \frac{rq_0}{\sqrt{(r^2-\omega^2)}}. \quad ...(iv)$$

From eq. (iii) and (iv), we get

$$A_1 = \frac{q_0}{2}\left(1+\frac{r}{\sqrt{(r^2-\omega^2)}}\right) \text{ and } A_2 = \frac{q_0}{2}\left(1-\frac{r}{\sqrt{(r^2-\omega^2)}}\right).$$

Putting these values in eq. (ii), we get

$$q = \frac{q_0}{2}\left(1+\frac{r}{\sqrt{(r^2-\omega^2)}}\right)e^{\left[-r+\sqrt{(r^2-\omega^2)}\right]t}$$

$$+\frac{q_0}{2}\left(1-\frac{r}{\sqrt{(r^2-\omega^2)}}\right)e^{\left[-r-\sqrt{(r^2-\omega^2)}\right]t}. \qquad ...(v)$$

There are three cases depending upon whether $r^2 >= \text{or } \omega^2$. Let us consider the case $r^2 < \omega^2$.

In this case $\sqrt{(r^2-\omega^2)}$ is an imaginary quantity. Let us put $\sqrt{(r^2-\omega^2)} = j\sqrt{(r^2-\omega^2)} = j\omega'$, where $\omega' = \sqrt{(\omega^2-r^2)}$.

The eq. (v) then becomes

$$q = \frac{q_0}{2}\left(1+\frac{r}{j\omega'}\right)e^{(-r+j\omega')t} + \frac{q_0}{2}\left(1-\frac{r}{j\omega'}\right)e^{(-r-j\omega')t}$$

$$= \frac{q_0}{2}e^{-rt}\left(e^{j\omega' t} + \frac{r}{j\omega'}e^{j\omega' t} + e^{-j\omega' t} - \frac{r}{j\omega'}e^{-j\omega' t}\right)$$

$$= \frac{q_0}{2}e^{-rt}\left(\frac{e^{j\omega' t}+e^{-j\omega' t}}{2} + \frac{r}{\omega'}\frac{e^{j\omega' t}-e^{-j\omega' t}}{2j}\right)$$

$$= \frac{q_0}{2}e^{-rt}\left(\cos\omega' t + \frac{r}{\omega'}\sin\omega' t\right)$$

Since $r << \omega$, we can ignore the term containing r/ω'. Then we have

$$\boxed{q = q_0\, e^{-rt}\cos\omega' t,}$$

where $\omega' = \sqrt{(\omega^2 - r^2)}$. Thus in this case the discharge is 'damped oscillatory'.

The condition for oscillatory discharge is $r^2 < \omega^2$, *i.e.*

$$\frac{R^2}{4L^2} < \frac{1}{LC}$$

or $$R < 2\sqrt{\left(\frac{L}{C}\right)}.$$

The period of discharge is

$$T = \frac{2\pi}{\omega'} = \frac{2\pi}{\sqrt{(\omega^2 - r^2)}} = \frac{2\pi}{\sqrt{\left(\frac{1}{LC} - \frac{R^2}{4L^2}\right)}}$$

The frequency of discharge is

$$n = \frac{1}{T} = \frac{1}{2\pi} = \sqrt{\left(\frac{1}{LC} - \frac{R^2}{4L^2}\right)}$$

$$= \frac{1}{2\pi}\sqrt{\left(\frac{1}{LC}\right)}, \text{ when R in small.}$$

Since $\frac{R^2}{4L^2}$ is a positive quantity, it is clear that in the presence of damping (when *R* is present) the frequency is less than in the absence of damping (when *R* is absent).

We note that the amplitude of oscillating charge is $q_0\, e^{-rt} = q_0\, e^{-\frac{R^2}{2L^2}t}$ which means that the amplitude decays exponentially with time. The decay is due to the resistance *R* in the circuit, because if *R=0*, the amplitude would remain constant at q_0. Thus in *L-C-R* circuit, the resistance is the only dissipative element.

Quality Factor: The quality factor Q of an oscillating system is defined as 2π times the ratio of the energy stored in the system to the energy lost per period. It may be shown to be given by

$$Q = \frac{2\pi\tau}{T}.$$

where *T* is the period of oscillation and τ is the relaxation time.

Q for L-C-R Circuit: The relaxation time for a damped oscillating system is given by

$$\tau = \frac{1}{2r}.$$

where r is damping constant. For *L-C-R* circuit, we have

$$2r = \frac{R}{L}.$$

$$\therefore \quad \tau = \frac{L}{R}.$$

The period T for the *L-C-R* circuit is given by

$$T = \frac{2\pi}{\sqrt{\left(\frac{1}{LC} - \frac{R^2}{4L^2}\right)}}$$

or when R is small,

$$T = 2\pi\sqrt{(LC)}$$

Therefore, for small R, the quality factor of an *L-C-R* circuit is given by

$$Q = \frac{2\pi\tau}{T} = \frac{1}{\sqrt{(LC)}}\,\frac{L}{R}$$

$$Q = \frac{1}{R}\sqrt{\left(\frac{L}{C}\right)}$$

This is the required expression.

Ballistic Galvanometer

A 'ballistic' galvanometer is a specially-designed galvanometer suitable for measuring the total *quantity of electricity*, or *charge*, displaced by a *varying current of short duration* such as in the charging or discharging of a capacitor and in electromagnetic induction. When such a transient current passes through the galvanometer, the moving part of the galvanometer experiences an angular impulse and begins to perform angular oscillations about its axis. The maximum deflection, or 'throw', obtained is proportional to the *total* charge which passed through the galvanometer.

Construction: A moving-coil ballistic galvanometer consists of a copper-wire coil of *large moment of inertia* and wound on a *non-conducting* frame of ivory or bamboo. The coil is suspended between cylindrical pole-pieces of a strong, laminated permanent magnet by means of a *long, thin* phosphor-bronze *strip.* The lower end of the coil is attached to a spring of phosphor-bronze wire. The charge enters at one terminal and after passing through the suspension, the coil, and the spring, leaves at the second terminal. A mirror is rigidly attached to the coil and the deflection of the coil is recorded by a lamp and scale arrangement.

As the coil has a large moment of inertia and its suspension has a small torsional constant, *its period of oscillation is fairly large.* Further, since the coil is wound on a *non-conducting* frame, *the electromagnetic damping is reduced.*

The damping due to viscosity of air is, however, present, but is usually *small.* These are the requirements for a galvanometer to be ballistic.

Equation of Motion of the Coil of B. G.: When the charge to be measured passes through the galvanometer, it gives an angular impulse to the coil which is set in angular oscillations.

Let I be the moment of inertia of the coil about its suspension and θ its angular displacement at any instant t. The instantaneous restoring torque acting on the coil is $-c\theta$ where c is the torsional constant of the suspension, and the instantaneous damping torque is $-b(d\theta/dt)$ where b is damping constant and $d\theta/dt$ the instantaneous angular velocity of the coil. Thus the total torque upon the coil is $-c\theta - b(d\theta/dt)$. This must be equal to the moment of inertia I multiplied by the instantaneous angular acceleration $d^2\theta/dt^2$. Thus

$$I\frac{d^2\theta}{dt^2} = -c\theta - b\frac{d\theta}{dt}$$

or

$$\frac{d^2\theta}{dt^2} + \frac{b}{I}\frac{d\theta}{dt} + \frac{c}{I}\theta = 0$$

Putting $b/I=2r$ and $c/I=\omega^2$, we get

$$\boxed{\frac{d^2\theta}{dt^2}+2r\frac{d\theta}{dt}+\omega^2\theta=0.} \qquad \text{...(i)}$$

This is exactly the differential equation of a damped harmonic oscillator. Hence the motion of the coil is damped oscillatory.

For small damping ($r<<\omega$). the solution of the above equation is of the form

$$\theta = \theta_0 e^{-rt} \sin\ (\omega' t+\phi),$$

where θ_0 and ϕ are constants, and $\omega' = \sqrt{(\omega^2-r^2)}$.

Applying the initial condition, $\theta=0$ at $t=0$, we get $\sin\phi=0$ or $\phi=0$. Further, for small r, $\omega'\simeq\omega$. Then we have

$$\theta = \theta_0 e^{-rt} \sin\ \omega t. \qquad \text{...(ii)}$$

Here $\theta_0 e^{-rt}$ represents the (damped) amplitude of the angular oscillations of the coil. θ_0 is the amplitude which would have been in the complete absence of damping. The period of oscillation is

$$T=\frac{2\pi}{\omega}=2\pi\sqrt{\left(\frac{I}{c}\right)} \qquad \text{...(iii)}$$

Relation Between Charge and Throw: Let us now find the value of θ_0 in terms of the charge passed through the galvanometer coil. If i be the current in the coil at an instant during the passage of the charge through it, then the instantaneous torque acting on the coil suspended in a radial magnetic field is Gi, where G is a galvanometer constant. This torque, in an infinitesimal time dt, gives to the coil an angular impulse (torque x time) $Gi\ dt$. Thus, if t' be the total time for which the current flows through the coil, the total angular impulse given to the coil is

$$\int_0^t Gi\,dt = G\int_0^t i\,dt = Gq,$$

where $q\left(=\int_0^t i\,dt\right)$ is the total charge passed through the coil.

Let us assume that the coil begins to rotate after the impulse is over. Then, if $(d\theta/dt)_0$ be its angular velocity *at start*, we have (angular momentum=angular impulse)

$$I\left(\frac{d\theta}{dt}\right)_0 = Gq$$

or
$$\left(\frac{d\theta}{dt}\right)_0 = \frac{Gq}{I}$$

Now, differentiating (ii), we get the instantaneous angular velocity of the coil. That is,

$$\frac{d\theta}{dt} = \theta_0 e^{-rt}\ \omega\ \cos\ \omega t - \theta_0 e^{-rt} r \sin\ \omega t.$$

At $t=0$, $\dfrac{d\theta}{dt} = \left(\dfrac{d\theta}{dt}\right)_0 = \dfrac{Gq}{I}$. Thus

$$\frac{Gq}{I} = \theta_0\omega$$

or
$$\theta_0 = \frac{Gq}{I\omega} = \frac{GqT}{2\pi I}. \qquad \left[\because\ \omega = \frac{2\pi}{T}\right]$$

But $T=2\pi\sqrt{(I/c)}$ so that $I=T^2c/4\pi^2$. Then

$$\theta_0 = \frac{2\pi Gq}{Tc}$$

or
$$q = \frac{T}{2\pi}\frac{c}{G}\ \theta_0 \qquad \text{...(iv)}$$

or $$q \propto \theta_0 .$$

Thus the charge passed through the galvanometer is proportional to the throw of the coil.

Logarithmic Decrement: Correction for Damping

In deducing the above relation it has been assumed that the damping is entirely absent. Actually it is not so. The motion of the coil is damped due to the viscosity of air. Therefore the coil oscillates with a *decreasing* amplitude. Hence the observed throw of the coil is smaller than its true value θ_0 which would have been if damping were entirely absent, A correction is thus necessary.

Let $\theta_1, \theta_2, \theta_3,$ ...be the successive throws *observed* at the end of the first, second, third........swings of the coil; θ_1, θ_3,...are on one side of the rest position of the coil and θ_2, θ_4... on the other. It is found that

$$\frac{\theta_1}{\theta_2} = \frac{\theta_2}{\theta_3} = \frac{\theta_3}{\theta_4} = ... = d,$$

where d is a constant which is called the 'decrement' per half oscillation, and $\log_e d$ is called the 'logarithmic decrement' λ. Therefore

$$\log_e . d = \lambda$$

or $$d = e^{\lambda}.$$

Thus, for half an oscillation, the decrement is

$$\frac{\theta_1}{\theta_2} = \frac{\theta_2}{\theta_3} = \frac{\theta_3}{\theta_4} = ...\, e^{\lambda},$$

and so for a full oscillation the decrement is

$$\frac{\theta_1}{\theta_2} = \frac{\theta_1}{\theta_3} \times \frac{\theta_2}{\theta_3} = e^{2\lambda},$$

and so on. Clearly, the decrement for a quarter of an oscillation will be $e^{\lambda/2}$.

To calculate the true throw θ_0 in the absence of damping, we must note that the *first* throw θ_1 is observed after the coil completes a quarter of an oscillation. Therefore, the decrement is given by

$$\frac{\theta_0}{\theta_1} = e^{\lambda/2}$$

or
$$\theta_0 = \theta_1 e^{\lambda/2}$$

$$= \theta_1 \left(1 + \frac{\lambda}{2} + \ldots\right),$$

as A is small, terms containing λ^2, λ^3, etc. can be neglected.

$$\therefore \quad \theta_0 = \theta_1 \left(1 + \frac{\lambda}{2}\right).$$

Substituting the value of θ_0 in eq. (iv), we get

$$q = \frac{T}{2\pi} \frac{c}{G} \theta_1 \left(1 + \frac{\lambda}{2}\right).$$

This is the relation between the charge passed and the first throw observed.

In practice, the value of λ is found by observing θ_1 and θ_{11}. Thus

$$\frac{\theta_1}{\theta_{11}} = e^{10\lambda}$$

or
$$\lambda = \frac{1}{10} \log_e \frac{\theta_1}{\theta_{11}} = \frac{1}{10} \times 2.3026 \times \log_{10} \frac{\theta_1}{\theta_{11}}$$

Difference between Dead-beat and Ballistic Galvanometers

A galvanometer is said to be 'dead-beat' if its coil, after deflection, quickly to the rest position without performing any oscillation. The conditions for this are :

The Period of Oscillation of the Coil should be small: For this the moment of inertia of the coil should be comparatively smaller and torsional constant of the suspension larger.

The Damping should be Heavy: For this the coil should be wound on a conducting frame and a soft-iron core be placed between the pole-pieces of the permanent magnet. This increases the electromagnetic damping.

On the other hand, a galvanometer is said to be ballistic if its coil makes a large number of oscillations before coming to rest, after the entire charge to be measured passes through it. The conditions for this are :

The Period of Oscillation of the Coil should be large: For this the moment of inertia of the coil should be large and torsional constant of the suspension small.

The Damping should be as small as possible: For this the coil should be wound on a non-conducting frame.

Electromagnetic Damping

Whenever a charge passes through the galvanometer, its coil rotates in the strong magnetic field of the permanent magnet. If the circuit of the coil is closed, an induced current is developed in the coil. This current, according to Lenz's law, produces a torque on the coil which opposes the motion of the coil. Further, if the coil is wound on a 'conducting' frame, then the frame forms a closed metallic circuit. Therefore, when the coil moves, induced current is set up in the frame also. This current also opposes the motion of the coil. Hence the coil soon comes to rest. This damping which arises due to the induced current in the moving system during Us motion in the permanent magnetic field, is called electromagnetic damping.

Short-circuiting of Galvanometer Terminals

When the terminals of a ballistic galvanometer, whose coil is oscillating, are joined together ; the circuit of the coil is closed and electromagnetic damping comes into play. Hence the coil stops oscillating.

Conversion of a Ballistic Galvanometer into a Deadheat One

For a galvanometer to be ballistic or dead-beat depends also on the external resistance of the circuit in which the galvanometer is placed. Smaller the resistance, larger will be the induced current and greater the damping. With a sufficiently small external resistance the motion ceases to be oscillatory and the galvanometer becomes over damped, that is, dead-beat.

Critical Damping

When the motion of the galvanometer coil is just non-oscillatory, that is, the deflected coil makes only one swing and returns slowly to its rest position, the galvanometer is said to be critically damped. The particular external resistance for which the motion just ceases to be oscillatory is called the 'critical external damping resistance'. With mote resistance the galvanometer is 'under-damped' and with less resistance it is 'overdamped'.

Grassot's Fluxmeter

It is an instrument for measuring the magnetic flux and hence the magnetic field directly. It is a special type of ballistic galvanometer. In it the restoring torque is made negligibly small and the electromagnetic damping large.

Construction: It consists of a coil *D* suspended by a single silk fibre (which has negligible torsional couple) and hanging between the pole-pieces *N* and S of a permanent magnet. *C* is the usual iron core. The upper end of the fibre is attached to a fiat spiral spring *E* to prevent damage from mechanical shocks. The current enters and leaves the coil by two fine silver spirals S_1 and S_2 which are connected to the terminals T_1 and T_2 The frame *FF* carries a pointer which moves over a scale.

A search coil (S.C.) of known cross-sectional area and number of turns is connected to the terminals T_1 and T_2.

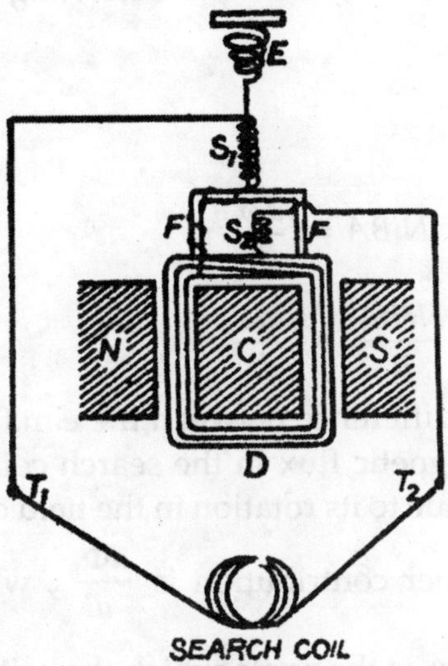

The search coil is placed in the space carrying the magnetic flux Φ_B to be measured and then withdrawn. The resulting deflection of the flux meter is noted. The magnetic flux is directly proportional to the deflection obtained.

Theory: When the search coil is withdrawn, there is a change in magnetic flux linked with it. An induced e.m.f. is set up in the circuit which sends a current through the coil D. Hence a deflecting couple acts upon the coil which therefore rotates. As the coil rotates, the magnetic flux through it due to the permanent magnet changes and an induced e.m.f. is set up in the circuit *in the opposite direction.* If be the *resultant* e.m.f. in the circuit at any instant and R the resistance of the circuit, then the current in the circuit at the instant is given by

$$i = \frac{e}{R}$$

Now, if N be the number of turns in the coil D, A its area of cross-section and B the magnetic field induction due to the permanent magnet, the deflecting couple at that instant is *NiBA*. If *I* be the moment of inertia of the coil and *w* its angular velocity at that instant, then the deflecting couple must be equal to $I\frac{d\omega}{dt}$.

That is,

$$NiBA = I\frac{d\omega}{dt}$$

or
$$\frac{NeBA}{R} = I\frac{d\omega}{dt}. \qquad ...(i)$$

Now *e* is the difference between the e.m.f. induced due to the change of magnetic flux in the search coil and that in the fluxmeter coil *D* due to its rotation in the field of the permanent magnet. The former contribution is $\frac{d\Phi_B}{dt}$, while the latter is $NBA\,\frac{d\theta}{dt}$, where θ is the angle which the coil makes from the mean position. Thus

$$e = \left(\frac{d\Phi_B}{dt} - NBA\frac{d\theta}{dt}\right).$$

Substituting for *e* in eq. (i), we have

$$\frac{NBA}{R}\left(\frac{d\Phi_B}{dt} - NBA\frac{d\theta}{dt}\right) = I\frac{d\omega}{dt}.$$

Integrating we have

$$\frac{NBA}{R}\int_0^t\left(\frac{d\Phi_B}{dt} - NBA\frac{d\theta}{dt}\right)dt = I\int_0^t\frac{d\omega}{dt}\,dt.$$

Now the integral on the right is $[\omega]_0^t$. Since the coil is at rest before and after the change in flux, ω =0 both at *t*=0 and at *t*=*t*. Hence the right-hand-side is zero. Thus

$$\int_0^t\left(\frac{d\Phi_B}{dt} - NBA\frac{d\theta}{dt}\right)dt = 0$$

or $$\int_0^{\Phi_B} d\Phi_B - NBA\int_0^{\theta} d\theta = 0$$

or $$\Phi_B - NBA\,\theta = 0$$

or $\Phi_B = NBA\,\theta = k\theta$ where k it is a constant.

Thus the magnetic flux is directly proportional to the deflection of the fluxmeter, and is independent of time during which the flux changes. The instrument can read magnetic flux directly by previous calibration. When the magnetic flux is known, the magnetic field can be evaluated, provided the area and the number of turns in the search coil are known.

Problems

1. A solenoid having a resistance of 5 ohm and self-inductance 10 henry is connected to a battery of 10 volts and negligible resistance. After how long will the current in it rise to 1 ampere ?

Solution: The growth of current in an *L-R* circuit is given by

$$i = i_0\,[1-e^{-(R/L)t}],$$

where i_0 $(=E/R)$ is the final steady current.

Here $i=1$ ampere, $i_0 = \frac{E}{R} = \frac{10\text{ volt}}{5\text{ ohm}} = 2$ amp and

$$\frac{R}{L} = \frac{5\text{ ohm}}{10\text{ henry}} = 0.5\text{ sec}^{-1}$$

$$\therefore\ 1 = 2\,(1-e^{-0.5t})$$

or $$\tfrac{1}{2} = 1-e^{-0.5t}$$

or $$e^{-0.5t} = 1-\tfrac{1}{2} = \tfrac{1}{2}$$

or $$e^{-0.5t} = 2$$

or $$0.5\,t = \log_e 2 = 0.69$$

$$\therefore\ t = \frac{0.69}{0.5} = 1.38\text{ sec.}$$

2. A l-μF capacitor is connected with an inductance of 10 in H and negligible resistance. Find the natural frequency of the electrical oscillations. If the maximum p.d. across the capacitor is 10 volt, calculate the energy of the oscillating circuit.

Solution: The natural frequency of *L*-C oscillations is given by

$$n = \frac{1}{2\pi}\sqrt{\left(\frac{1}{LC}\right)}.$$

Here $L=10$ mH$=10 \times 10^{-3}$ H and $C=1\mu F=1 \times 10^{-6}$ F.

$$\therefore\ n = \frac{1}{2\times3.14}\sqrt{\left(\frac{1}{(10\times10^{-3})(10\times10^{-6})}\right)}$$

$$= \frac{10^4}{2\times3.14} = 1592\ \text{sec}^{-1}.$$

The energy of the oscillating circuit

$$= \tfrac{1}{2}\ CV^2 = \tfrac{1}{2}\ (1 \times\ 10^{-6}\ \text{farad})\ (10\ \text{volt})^2$$

$$=5\text{x}10^{-5}\ \text{joule}.$$

3. If a 10-pF capacitor is charged to a potential of 10,000 V and then discharged through an inductance of 0.001 H, calculate the amplitude of current oscillations.

Solution: The discharge of a capacitor C through an inductance L is given by

$$q = q_0 \cos \omega t,$$

where $\omega = 1\sqrt{(LC)}$. The instantaneous current is given by

$$i = \frac{dq}{dt} = -\ q_0\,\omega \sin\ \omega t.$$

The current amplitude is

$$i_0 = q_0\,\omega.$$

But $q_0 = CV$ and $\omega = 1\sqrt{(LC)}$.

$$\therefore i_0 = \frac{CV}{\sqrt{(LC)}} = V\sqrt{\left(\frac{C}{L}\right)}$$

$$=(10,000 \text{ volt})\sqrt{\left(\frac{10\times10^{-6}\text{ farad}}{0.001\text{ henry}}\right)}$$

$$= 1000 \text{ amp}$$

4. A capacitor of 20μF is charged to 1500 volts and then discharged through a coil of negligible resistance and 0.002 henry inductance. Calculate the frequency and maximum current.

[Ans. 796 sec^{-1}, 150 amp.]

5. A capacitor of 0.1μF, an inductance of 0.1 henry and a resistance of 800 ohm are connected in series. This combination is connected with a battery. Is the current in the circuit oscillatory?

Solution: The current is oscillatory, if

$$\frac{R^2}{4L^2} < \frac{1}{LC}$$

or $$R < 2\sqrt{\left(\frac{L}{C}\right)}$$

Here $$2\sqrt{\left(\frac{L}{C}\right)} = 2\sqrt{\left(\frac{0.1\text{ henry}}{0.1\times10^{-6}\text{ farad}}\right)} = 2000 \text{ ohm.}$$

The resistance R(=800 ohm) is less than this. Hence the current is oscillatory.

6. A circuit has 0.4 henry inductance and 0.0024 uF capacitance. How much resistance should be added to make it oscillatory?

Solution: The circuit is oscillatory, if

or $$R < 2\sqrt{\left(\frac{L}{C}\right)}$$

Here $2\sqrt{\left(\frac{L}{C}\right)} = 2\sqrt{\left(\frac{0.4 \text{ henry}}{0.0024\times10^{-6}\text{ farad}}\right)}$ $=2.58\times10^{4}$ ohm.

Hence the resistance to be added should be less than 2.58×10^{4} ohm.

7. An L-C-R circuit having L = 0.2 henry and R=100 ohm is critically damped. What is the value of C ?

Solution: The condition of critical damping is

$$\frac{R^2}{4L^2} = \frac{1}{LC}$$

or
$$C = \frac{4L}{R^2}.$$

Putting the given values :

$$C = \frac{4\times0.2 \text{ henry}}{(100 \text{ ohm})^2} = 8\times10^{-5} \text{ farad} = 80\mu\text{F}.$$

8. A condenser of 5μF capacity is discharged through 0.2 ohm resistance and 2mH inductance. Calculate the frequency and Q-factor of the circuit.

Solution: The frequency of oscillations, at small resistance, is

$$n = \frac{1}{2\pi}\sqrt{\left(\frac{1}{LC}\right)}$$

Here L=2 mH=2×10^{-3} henry and C=5μF=5×10^{-6}farad.

$$n = \frac{1}{2\times3.14}\sqrt{\left(\frac{1}{2\times10^{-3}\times5\times10^{-6}}\right)} = \frac{10^2}{2\times3.14} = 1592 \text{ cps.}$$

The quality factor of the circuit is given by

$$Q = \frac{1}{R}\sqrt{\left(\frac{L}{C}\right)}$$

$$= \frac{1}{0.2}\sqrt{\left(\frac{2\times10^{-3}}{5\times10^{-6}}\right)}$$

$$= \frac{1}{0.2} \times 20 = 100.$$

9. Show that for a circuit with L=10 millihenry, C=lμF and R=0.1 ohm, the amplitude of the charge oscillations will be reduced to half in 0.14 sec. ($\log_e 2$=0.69). What is the quality factor of the circuit?

Solution: The equation of the damped oscillatory discharge is

$$q = q_0 e^{-rt} \cos \omega' t,$$

where $r=R/2L$ and $\omega' = \sqrt{\left(\frac{1}{LC} - \frac{R^2}{4L^2}\right)}$.

The (damped) amplitude at any instant t is $q_0\, e^{-rt}$, where q_0 is the initial (at t=0) amplitude.

Now, if $q_0/2$ be the amplitude at time t, then we have

$$\frac{q_0}{2} = q_0 e^{-rt}$$

or $$2 = e^{rt}.$$

or $$rt = \log_e 2 = 0.69$$

$$\therefore \quad t = \frac{0.69}{r} = 0.69 \times \frac{2L}{R}$$

$$= 0.69 \times \frac{2 \times (10 \times 10^{-3}\,\text{henry})}{0.1\,\text{ohm}}$$

$$= 00.14 \text{ sec.}$$

The quality factor of the circuit is given by

$$Q = \frac{1}{R}\sqrt{\left(\frac{L}{C}\right)}$$

$$= \frac{1}{0.2}\sqrt{\left(\frac{2 \times 10^{-3}}{2 \times 10^{-6}}\right)}$$

$$= \frac{1}{0.2} \times 20 = 1000.$$

10. A moving-coil galvanometer has a periodic time 2 π sec and restoring couple 900 dyne-cm/rad. If the amplitude decreases to 1/ 10 of its original value in 69 sec, find the damping constant and the quality factor. ($\log_e 10 = 2.3$).

Solution: The instantaneous (damped) amplitude of an oscillating galvanometer coil is $\theta_0 e^{-rt}$ where θ_0 is the initial (at $t=0$) or undamped amplitude, and $r = b/2I$, where b is damping constant.

Now, if $\frac{\theta_0}{10}$ be the amplitude at t=69 sec, then we have

$$\frac{\theta_0}{10} = \theta_0 e^{-69r}$$

$$r = \frac{1}{69} \log_e 10 = \frac{1}{69} \times 2.3 = \frac{1}{30} \text{ sec}^{-1}.$$

Now, the period of coil oscillations is given by

$$T = 2\pi\sqrt{\left(\frac{I}{c}\right)}$$

where I is the moment of inertia of the coil and c is the restoring couple per unit twist. From this, we have

$$I = \frac{T^2 c}{4\pi^2}$$

$$= \frac{(2\pi \text{ sec})^2 (900 \text{ dyne-cm})}{4\pi^2}$$

$$= 900 \text{ gm-cm}^2.$$

The damping constant is therefore

$$b = 2Ir$$

$$= 2 \times 900 \text{ gm-cm}^2 \times \frac{1}{30} = \text{sec}^{-1}$$

$$= 60 \text{ gm-cm}^2/\text{sec}.$$

The quality factor of the oscillating system is given by

$$Q = \frac{2\pi\tau}{T}, \text{ where } \tau \text{ is relaxation time}$$

$$= \frac{2\pi}{T}\left(\frac{1}{2r}\right) \qquad \left[\therefore \ \tau = \frac{1}{2r}\right]$$

$$= \frac{2\pi}{2\pi \text{ sec}} \ \frac{1}{2 \times \frac{1}{30} \text{sec}^{-1}}$$

$$= 15.$$

11. A 0.2-pF capacitor is charged to 3 volts. When discharged through a ballistic galvanometer, it gives a corrected deflection of 10 cm. If the period of the galvanometer is 10 sec, calculate its current sensitivity.

Solution: The charged passed through the galvanometer is

$$q = CV$$

$$= (0.2 \ \mu\text{F}) \ (3 \text{ volt}) = 0.6 \ \mu\text{C}.$$

$$\therefore \ \text{charge sensitivity} = \frac{\text{deflection in mm}}{\text{charge in micro-coulomb}}$$

$$= \frac{100}{0.6} \ \text{mm}/\mu\text{C}$$

and

$$\text{current sensitivity} = \frac{T}{2\pi} \times \text{charge sensitivity}$$

$$= \frac{10 \text{ sec}}{2 \times .314} \times \frac{100}{0.6} \times \frac{\text{mm}}{\text{micro-coul}}$$

$$= 265 \text{ mm/micro-amp}.$$

7

Forced Oscillations

Free and Forced Oscillations

Free Oscillations: When a body capable of oscillation is displaced from its equilibrium position and released, it begins to oscillate. The frequency of oscillations depends upon certain properties of the body. Such oscillations are called 'free oscillations' and their frequency is called the 'natural frequency' of the body. Usually the amplitude of such oscillations gradually decreases to zero owing to the presence of certain frictional (damping) forces. Hence they are also called as 'damped oscillations'.

Forced Oscillations: When a body is set in oscillation under an external *periodic* force then, in the beginning, the body tries to oscillate with its own natural frequency, but very soon these oscillations die out and the body oscillates with the frequency of the external applied force. Such oscillations are called 'forced oscillations'. The amplitude of the forced oscillations depends upon the relation between the frequency of the applied force and the natural frequency of the body.

When a tuning fork is struck and its stem is placed on a

table, the table is set in oscillations with the frequency of the fork. Similarly, when an alternating current is passed through a stretched string held between the poles of a magnet, the string oscillates with the frequency of the current. These oscillations of the string are the forced oscillations.

Resonance

When a body is set into oscillation by an external periodic force *of the same frequency as the natural frequency of the body,* the amplitude of oscillation is very much increased. This phenomenon is called "resonance". Thus the resonance is a special case of forced oscillation.

The apparatus consists of a heavy pendulum A and three light pendulums B,C,D of different lengths, all suspended from the same cord. The pendulum A is set into oscillation. Soon after the pendulums B, C and D receive periodic impulses at the frequency of A through the cord and begin to oscillate.

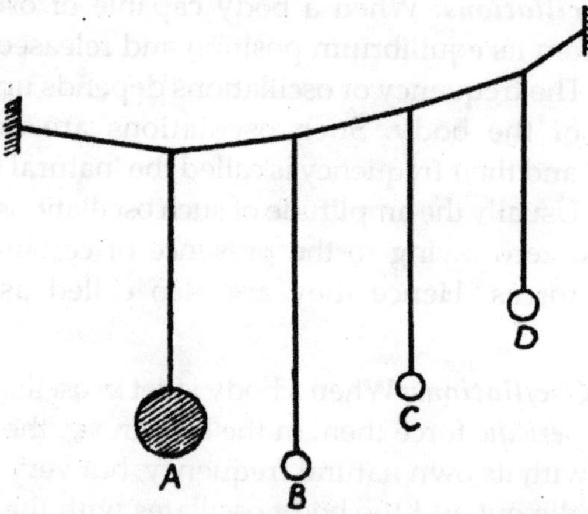

This is an example of forced oscillations. It is seen that *B* and *D,* whose lengths (and hence natural frequencies) are different from that of *A,* have a small amplitude. But the pendulum C, whose length (and so natural frequency) is same as that of *A,* oscillates with a much larger amplitude. This is the case of resonance.

The explanation of resonance is very simple. When the frequency of the external force is equal to the natural frequency of the body then at each step the force helps the body in going ahead in its oscillation. The successive impulses given by the periodic force are added up and *increase the amplitude of oscillation of the body continuously.* The loss of energy of the body due to air resistance and internal friction also increases with the increase of its amplitude of oscillation. Finally, a stage is reached when the energy supplied by the external force becomes equal to the energy lost by the body. This condition of equilibrium is reached a little before the amplitude becomes very large. If there were no loss of energy, the amplitude would have become infinite.

Examples

Mechanical Resonance: (i) When a tuning fork is struck and its stem is placed upon a hollow box of dimensions such that the air inside the box has a natural frequency equal to that of the fork, an intense note is heard.

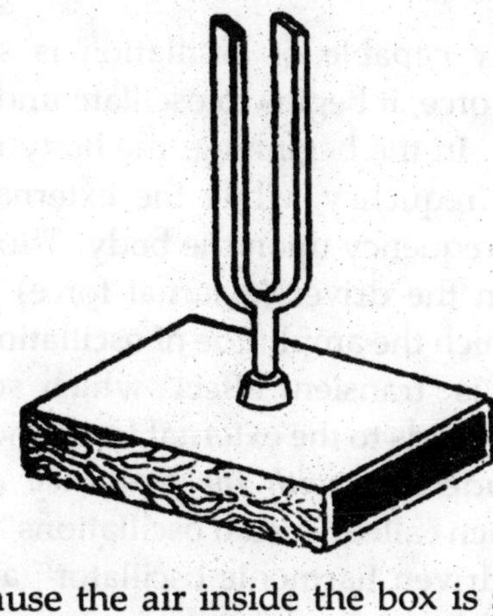

This is because the air inside the box is set into resonant oscillations of large amplitude.

(ii) If two springs stretched upon the same board have the same frequency and one is set in oscillation, the other begins to oscillate.

(iii) Objects placed on a piano may be set in resonant oscillations by the sounding of a particular note.

(iv) Soldiers marching over a suspension bridge always break steps because if the frequency of their march incidently coincides with the natural frequency of the bridge, the bridge may be set into violent oscillations and collapse.

(v) *Electrical Resonance:* A tuned circuit in a radio receiver catches only those frequencies which are very near its own natural frequency. This is why it is possible to listen to a particular station.

(vi) *Optical Resonance:* When an atom or molecule is struck by electromagnetic radiation whose frequency is exactly equal to that of one of the natural vibrations of the atom or molecule, the radiation is absorbed more readily than at other frequencies. This is 'resonance absorption' of radiation.

Forced Oscillations

When a body capable of oscillation is subjected to an external *periodic* force, it begins to oscillate under the action of the applied force. In the beginning, the body tries to oscillate with its natural frequency, while the external force tries to impose its own frequency upon the body. Thus there is a sort of tussle between the driver (external force) and the driven (body) during which the amplitude of oscillation rises and falls alternately. This is 'transient effect' which soon dies away. Finally, the body yields to the external force and oscillates with *constant* amplitude and *with the frequency of the force.* Its oscillations are then called "forced oscillations". The oscillating body is called "driven harmonic oscillator" and the external force is called "driving force."

Equation of Forced Oscillations: Let us consider a system oscillating about an equilibrium position under an external *periodic* force. Let x be its displacement from the equilibrium position at an instant during the oscillation. Its instantaneous velocity is dx/dt. The forces acting upon the system at this instant are:

(i) A restoring force proportional to the displacement but acting in the opposite direction. This may be written as

$$-kx,$$

where k is the force-constant.

(ii) A frictional force proportional to the velocity but acting in the opposite direction. This may be written as

$$-b\frac{dx}{dt}$$

where b is a positive constant.

(iii) An external periodic force represented by

$$F_0 \sin pt,$$

where F_0 is the maximum value of this force and p is its angular frequency. Thus the total force F acting upon the system is

$$F= -kx-b\frac{dx}{dt} +F_0 \sin pt.$$

By Newton's second law this must be equal to the product of the mass m of the system and the instantaneous acceleration $\frac{d^2x}{dt^2}$

That is

$$-kx-b\frac{dx}{dt} +F_0 \sin pt=m \frac{d^2x}{dt^2}$$

or

$$\frac{d^2x}{dt^2}+\frac{b}{m}\frac{dx}{dt}+\frac{k}{m}x=\frac{F_0}{m}\sin pt.$$

Putting $\frac{b}{m}= 2r$, $\frac{k}{m}=\omega^2$ *and* $\frac{F_0}{m}=f_0$ we get

$$\boxed{\frac{d^2x}{dt^2}+2r\frac{dx}{dt}+\omega^2x=f_0 \sin pt.} \qquad ...(i)$$

This is the differential equation of motion of the forced harmonic oscillator.

Once the transient effect dies away and a steady state is set up, the system will oscillate with the frequency of the applied force. This is a necessary physical condition of the oscillations. Let us, therefore, try the following equation as a solution of eq. (i) :

$$x = A \sin (pt-\theta), \qquad :(ii)$$

where A and θ are arbitrary constants.

Differentiating eq. (ii) twice with respect to t, we get

$$\frac{dx}{dt} = pA \cos (pt-\theta).$$

and $$\frac{d^2x}{dt^2} = p^2A \sin (pt-\theta) .$$

Substituting the values of $x, \frac{dx}{dt}$ and $\frac{d^2x}{dt^2}$ in eq. (i), we get

$$-p^2 A \sin (pt-\theta) + 2 rpA \cos (pt-\theta)+\omega^2 A \sin (pt-\theta)$$
$$=f_0 \sin \{(pt-\theta)+\theta\}$$

or $$A(\omega^2-p^2) \sin (pt-\theta) +2 rpA \cos (pt-\theta)$$
$$=f_0 \sin (pt-\theta) \cos \theta +f_0 \cos (pt-\theta) \sin \theta .$$

If this equation is to be satisfied for all values of t, then the coefficients of sin $(pt-\theta)$ and cos $(pt-\theta)$ on the two sides must be equal Equating them, we obtain

$$A (\omega^2-p^2) =f_0 \cos\theta \qquad ...(iii)$$

and $$2 rpA=f_0 \sin\theta . \qquad ...(iv)$$

Squaring and adding eq. (iii) and (iv), we get

$$A^2\{(\omega^2 - p^2)^2 + 4r^2p^2\} = f_0^2$$

or
$$A = \frac{f_0}{\sqrt{\left[(\omega^2 - p^2)^2 + 4r^2p^2\right]}}. \qquad \text{...(V)}$$

Dividing eq. (iv) by eq. (iii), we get

$$\tan\theta = \frac{2rp}{\omega^2 - p^2}. \qquad \text{...(vi)}$$

Substituting the value of A from eq. (v) in eq. (ii), we get

$$x = \frac{f_0}{\sqrt{\left[(\omega^2 - p^2)^2 + 4r^2p^2\right]}}\sin(pt - \theta) \qquad \text{...(vii)}$$

where θ is given by eq. (vi). This is the solution of the differential equation of the forced harmonic oscillator. This shows that the oscillator has a constant amplitude A and a frequency equal to that of the impressed force, and lags behind the force in phase by θ.

Amplitude Resonance: Eq. (v) shows that the amplitude of the forced oscillations depends upon $(\omega^2 - p^2)$ which in turn depends upon the difference between the driving frequency p and the natural frequency ω of the oscillator. Smaller this difference, larger the amplitude.

At very low driving frequency $(p << \omega)$, we have

$$A \simeq \frac{f_0}{\omega^2} \simeq \frac{F_0/m}{k/m} \simeq \frac{F_0}{k},$$

which shows that the amplitude depends only on the force-constant.

It is independent of the mass, of the, damping and of the driving frequency.

At very high driving frequency ($p >> \omega$), we have

$$A \simeq \frac{f_0}{p^2} \simeq \frac{F_0}{mp^2},$$

which shows that the amplitude, which now depends upon the mass, continuously decreases as the driving frequency p is further increased.

At driving frequencies comparable with the natural frequency, the amplitude is maximum for a particular frequency. The phenomenon of amplitude becoming a maximum is known as "amplitude resonance", and the particular driving frequency is known as the "resonant frequency".

Now the amplitude will be a maximum when the denominator $\sqrt{\left[(\omega^2 - p^2)^2 + 4r^2p^2\right]}$ in eq. (v) is a minimum. This will be so when its first differential coefficient is zero, *i.e.* when

$$\frac{d}{dp}\left[(\omega^2 - p^2)^2 + 4r^2p^2\right] = 0$$

or $$2(\omega^2 - p^2)\ (-2p) + 4\ r^2(2\ p) = 0$$

or $$\omega^2 - p^2 = 2\ r^2$$

$$p = \sqrt{(\omega^2 - 2r^2)}. \qquad \text{...(viii)}$$

Thus the amplitude of forced oscillations is maximum when the frequency $p/2\pi$ it of the impressed force is

$$\frac{\sqrt{(\omega^2 - 2r^2)}}{2\pi} \qquad \text{(resonant frequency)}$$

which is slightly less than the natural frequency $\omega/2\pi$ as well as than the damped frequency $\frac{\sqrt{(\omega^2 - r^2)}}{2\pi}$ of the system. If, however, the damping is small (*r* is small), the resonant frequency is very nearly equal to the natural frequency $\omega/2\pi$. In the ideal case (r=0), eq. (viii) becomes

$$p = \omega,$$

i.e. the amplitude resonance occurs when the frequency of the impressed force coincides with the natural frequency of the oscillator. This is the condition of resonance in the absence of damping.

Effect of Damping on the Response at Resonance: Substituting the condition of maximum amplitude, $p=\sqrt{(\omega^2-2r^2)}$ in eq. (v), we get

$$A_{max}=\frac{f_0}{2r\sqrt{(r^2+p^2)}} \qquad ...(ix)$$

This shows that the maximum amplitude depends upon the damping, ***smaller the damping, larger the maximum amplitude.*** In the ideal case (r=0), $A_{max} \to \infty$. Thus we can see that damping controls the response at resonance.

Sharpness of Resonance

The amplitude of forced oscillations is a maximum when the frequency of the applied force has a value which satisfies the condition of resonance. As soon as the frequency changes from this value, the amplitude falls. If the fall in amplitude for a small departure of the frequency from the resonant value is considerable, the resonance is said to be "sharp". If, on the other hand, the fall is small, the resonance is said to be "flat".

Effect of Damping on the Sharpness of Resonance: The amplitude of forced oscillations is given by

$$A=\frac{f_0}{\sqrt{\left[(\omega^2-r^2)^2+r^2p^2\right]}}$$

where w is the natural frequency of the oscillates, *p* is the frequency of the applied force and r *is* the damping factor". This shows that the amplitude of forced oscillation depends upon the relative magnitude of the frequencies p and w, and upon the damping r. In the figure below, the amplitude of forced oscillation, A, has been plotted against the ratio p/ω *for* various dampings.

For very low driving frequency p, the amplitude is nearly same for all values of damping.

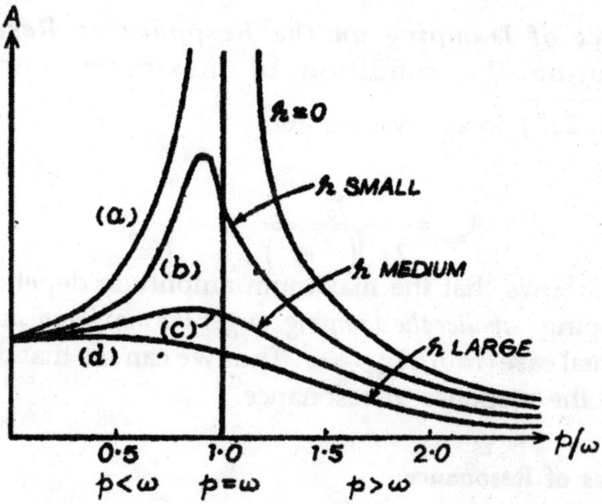

As p increases, the amplitude increases and becomes maximum at a certain value of p, which depends on damping. Curve (a) shows the amplitude when r=0, that is, when there is no damping. In this case the amplitude becomes infinite *at* $p=\omega$. Curve (b), (c) and (d) show that as damping (r) increases, the peak of the curve moves towards the left, *i.e.* the value of p for which the amplitude is a maximum decreases. Further, as damping (r) increases, the peak moves downward, *i.e.* the maximum amplitude of the forced oscillation is lowered.

As p further increases, the amplitude again falls and tends towards zero whatever be the damping.

It is seen that the curve for small r falls much more rapidly on either side of the peak than those for larger r. This means that for the same departure from resonance condition, the amplitude of oscillation falls rapidly when the damping is small, and slowly when the damping is large. In other words, *smaller the damping, sharper the resonance.*

Example: A sonometer wire resounds to a vibrating tuning fork only when the tuning is exact, whereas a column of air resounds even when imperfectly tuned. An example of sharp resonance is found in case of sonometer wire in which the damping is small. When the wire is tuned exactly to a given tuning fork, it oscillates vigorously but even for a very slight change from the correct length the oscillations practically vanish.

An example of flat resonance is found in case of a resonance column in which the oscillations of the air are subjected to a larger damping. When the length of the air-column is adjusted to match its frequency with the frequency of a given tuning fork, the column responds over a wide range and it is difficult to find out the exact resonant length of the column.

Velocity Amplitude of Forced (Driven) Oscillations

When a system of mass m having a natural angular frequency w, oscillates under an external force F_0 sin pt, the displacement of the forced oscillations at any instant is given by

$$x=\frac{f_0}{\sqrt{\left[\left(\omega^2-r^2\right)^2+4r^2p^2\right]}}\sin(pt-\theta), \qquad \text{...(i)}$$

where f_0=(F_0/m), ω^2 -=$\frac{k}{m}$ where k is spring factor, r (=$b/2m$) is a damping factor and $\theta=\tan^{-1}\frac{2rp}{\omega^2-p^2}$ Therefore, the instantaneous velocity is

$$u_A=\frac{dx}{dt}=\frac{f_0 p}{\sqrt{\left[\left(\omega^2-p^2\right)^2+4r^2p^2\right]}}\cos(pt-\theta). \qquad \text{...(ii)}$$

The velocity will be maximum when cos $(pt-\theta)=1$. The maximum value is known as "velocity-amplitude" u_A. Thus

$$u_A = \frac{f_0 p}{\sqrt{\left[(\omega^2 - p^2)^2 + 4r^2p^2\right]}}. \qquad \text{...(iii)}$$

This is the expression for the velocity-amplitude of a driven harmonic oscillator.

Eq. (iii) shows that the velocity-amplitude depends upon the value of the driving frequency p relative to the natural frequency ω.

At very low driving frequencies ($p \ll \omega$), we may write

$$u_A \simeq \frac{f_0 p}{\omega^2} \simeq \frac{(F_0/m)\,p}{k/m} \simeq \frac{f_0 p}{k}$$

which shows that the amplitude is mainly governed by the force-constant k (spring factor).

At very high driving frequencies ($p \gg \omega$), we may write

$$u_A \simeq \frac{f_0}{p} \simeq \frac{F_0}{mp},$$

which shows that the amplitude is mainly governed by the mass m (inertia factor).

Velocity Resonance: At driving frequencies comparable with natural frequency, the velocity amplitude is maximum for a particular frequency. For this let us write eq. (iii) as

$$u_A = \frac{f_0}{\sqrt{\left[\left(\frac{\omega^2 - p^2}{p}\right)^2 + 4r^2\right]}}$$

This is maximum when the denominator is minimum, *i.e.* when

$$\left(\frac{\omega^2 - p^2}{p}\right)^2 = 0$$

$$p = \omega .$$

Thus, whatever be the damping, the velocity-amplitude is maximum when the driving frequency is fual to the natural undamped frequency of the oscillator. This phenomenon is known as 'velocity resonance'.

The velocity-amplitude at resonance is $f_0/2r = F_0/b$, which shows that it is governed only by damping.

Difference between Amplitude Resonance and Velocity Resonance. The amplitude resonance occurs when the driving frequency is slightly lower than the natural frequency of the oscillator, depending upon, the magnitude of damping. The velocity resonance occurs when the two frequencies are equal, whatever the damping.

At amplitude resonance the displacement lags behind the driving force by $\pi/2$, while the velocity at resonance is always in phase with the driving force.

Phase of the Displacement and Velocity of Forced Oscillations

When a system of mass m having a natural angular frequency ω, oscillates under an external force F_0 *sin pt*, the instantaneous displacement and velocity of the system are given by

$$x = \frac{f_0}{\sqrt{\left[\left(\omega^2 - p^2\right)^2 + 4r^2p^2\right]}} \sin(pt - \theta), \qquad \text{...(i)}$$

and $$u = \frac{dx}{dt} = \frac{f_0 p}{\sqrt{\left[\left(\omega^2 - p^2\right)^2 + 4r^2p^2\right]}} \cos(pt - \theta)$$

or $$u = \frac{f_0 p}{\sqrt{\left[\left(\omega^2 - p^2\right)^2 + 4r^2p^2\right]}} \sin\left\{pt - \left(\theta - \frac{\pi}{2}\right)\right\}, \qquad \text{...(ii)}$$

where $f_0 = (F_0/m)$, $\omega^2 = k/m$ where k is spring factor, $r(=b/2m)$ is a damping factor and

$$\boxed{\tan\theta = \frac{2rp}{\omega^2 - p^2}} \qquad \text{...(iii)}$$

Thus the displacement of the forced oscillations, given by eq. (i), lags behind the driving force F_0 sin pt by an angle θ given by eq. (iii). This shows that the phase difference θ depends upon the damping r, and upon the difference between p and ω. Let us consider the following cases :

If $p<\omega$, tan θ is positive, θ lies between θ and $\pi/2$.

If $p=\omega$, tan θ is infinite, $\theta = \pi/2$.

If $p>\omega$, tan θ is negative, θ lies between $\pi/2$ and π.

Thus the displacement of the forced oscillations always lags behind the driving force in phase by an angle between 0 and π, depending upon the value of p relative to ω. If damping is very small ($r \simeq 0$), then

$$\tan\theta = \frac{2rp}{\omega^2 - p^2} \simeq 0.$$

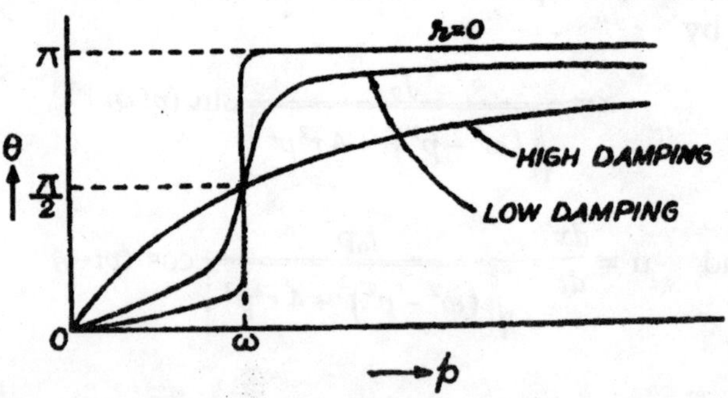

Thus, when $p<\omega$, tan θ is very small but positive, so that θ is nearly zero. This means that under this condition the forced oscillations are practically in phase with the driving force.

When $p > \omega$, tan θ is very small but now negative, so that θ is very nearly π. Under this condition the forced oscillations are almost opposite in phase with the driving force.

When $p = \omega$, tan $\theta = \infty$ so that $\theta = \pi/2$. Thus at resonance, the oscillations always lag behind the force by $\pi/2$. It means that at resonance the displacement is minimum when the driving force is maximum and vice-versa.

Let us now consider the phase behaviour of the velocity of the oscillator. According to eq. (ii), the velocity lags behind the driving force F_0 sin *pt* by an angle ϕ, (say), where

$$\phi = \theta - \frac{\pi}{2}.$$

when $p < \omega$, tan θ is positive *i.e.* θ lies between 0 and $\pi/2$, so that ϕ is negative. This means that the velocity *leads* the driving

When $P > \omega$, tan θ is negative, *i.e.* θ lies between $\pi/2$ and π, so that ϕ is positive. This means that the velocity *lags* behind the driving force.

When $p = \omega$ (at velocity resonance), tan $\theta = \infty$, so that

$$\theta = \frac{\pi}{2}$$

and
$$\phi = \frac{\pi}{2} - \frac{\pi}{2} = 0$$

Thus at resonance the velocity is always in phase with the driving force.

Average Power Absorbed by Driven Oscillator

A driven oscillator absorbs energy from the driving force

which is dissipated in doing work against the damping forces present.

Let m be the mass of the oscillator whose natural angular frequency is w, and F_0 sin pt the external driving force. Then, *in the steady state*, the displacement of the oscillator at any instant t is given by

$$x = \frac{f_0}{\sqrt{\left[\left(\omega^2 - p^2\right)^2 + 4r^2p^2\right]}} \sin(pt - \theta).$$

where $f_0 = \frac{F_0}{m}$ $r = \frac{b}{2m}$ (*b* being the damping constant),

$\omega = \sqrt{(k/m)}$ (*k* being the spring factor), and $\tan\theta = \dfrac{2rp}{\omega^2 - p^2}$. The instantaneous velocity is

$$\frac{dx}{dt} = \frac{f_0 p}{\sqrt{\left[\left(\omega^2 - p^2\right)^2 + 4r^2p^2\right]}} \cos(pt - \theta).$$

The instantaneous power P absorbed by the oscillator is equal to the product of the instantaneous driving force and the instantaneous velocity. That is

$$P = (F_0 \sin pt)\frac{dx}{dt}$$

$$= \frac{mf_0^2 p}{\sqrt{\left[\left(\omega^2 - p^2\right)^2 + 4r^2p^2\right]}} \sin pt \cos(pt - \theta)$$

$$= \frac{mf_0^2 p}{\sqrt{\left[\left(\omega^2 - p^2\right)^2 + 4r^2p^2\right]}} (\sin pt \cos pt \cos\theta + \sin^2 pt \sin\theta).$$

Let us calculate the average values of the periodic terms for one period T $(=2n/p)$. We have

$$\frac{1}{T}\int_0^T \sin pt \cos pt \, dt = 0$$

and $$\frac{1}{T}\int_0^T \sin^2 pt\, dt = \tfrac{1}{2}$$

Hence the average power absorbed is

$$P_{av} = \frac{mf_0^{\,2}p}{\sqrt{\left[\left(\omega^2-p^2\right)^2+4r^2p^2\right]}}\left(\tfrac{1}{2}\sin\theta\right),$$

Since $\tan\theta = \dfrac{2rp}{\omega^2-p^2}$, we have $\sin\theta = \dfrac{2rp}{\sqrt{\left[\left(\omega^2-p^2\right)^2+4r^2p^2\right]}}$

Therefore

$$\boxed{P_{av} = \frac{mf_0^{\,2}rp^2}{\left(\omega^2-p^2\right)^2+4r^2p^2}.}$$

This is the expression for the average power supplied by the driving force.

Power Dissipation by Driven Oscillator: The power absorbed from the driving force is dissipated in doing work against the damping force *–b (dx/dt) or — 2mr (dx/dt).*

The rate of work (instantaneous power) *P′* against the damping force is given by (force × velocity)

$$P' = \left(2mr\frac{dx}{dt}\right)\left(\frac{dx}{dt}\right)$$

$$= 2mr\left(\frac{dx}{dt}\right)^2$$

$$= 2mr\,\frac{f_0^{\,2}p}{\left(\omega^2-p^2\right)^2+4r^2p^2}\cos^2(pt-\theta).$$

Now, the average of $\cos^2(pt-\theta)$ for one period *T*,

i.e. $\frac{1}{T}\int_0^T \cos^2(pt-\theta)\,dt = \frac{1}{2}$. Therefore, the average power dissipated is

$$\boxed{P'_{av} = \frac{f_0^{\,2} r p^2}{\left(\omega^2 - p^2\right)^2 + 4r^2p^2}.}$$

This is same as P_{av} (the average power absorbed), Thus, in the steady state, the power supplied by the driving force equals the power dissipated.

Maximum Power Absorption

The average power absorbed is given by

$$P_{av} = \frac{m f_0^{\,2} r p^2}{\left(\omega^2 - p^2\right)^2 + 4r^2p^2}.$$

It is maximum when the denominator $\left(\omega^2 - p^2\right)^2 + 4r^2p^2$ is a minimum. The minimum value of a squared term is zero. Since neither r nor p can be zero, therefore $4\,r^2p^2$ cannot be zero. Thus for maximum power we must have

$$\left(\omega^2 - p^2\right)^2 = 0$$

or $$P = \omega,$$

which is the condition of 'velocity resonance'. Hence we conclude that the power transferred from the driving force to the oscillator is maximum at the frequency of velocity-resonance (not at the frequency of amplitude resonance.

$$Pav\text{ (maximum)} = \frac{m f_0^{\,2} r p^2}{4r^2p^2} = \frac{m f_0^{\,2}}{4r}.$$

Sharpness of Resonance

The amplitude of forced oscillations is maximum for a

particular value of the driving frequency. For small damping, this value of the driving frequency is almost equal to the natural frequency of the oscillator. Under this condition, the oscillator is said to be in "resonance".

As the driving frequency deviates either way from its value for resonance, the amplitude of oscillations falls from its maximum value. If the amplitude falls rapidly with change in driving frequency, the resonance is said to be "sharp".

Band-width of Resonance: The forced oscillator absorbs power from the driving force, and the average power absorbed depends upon the value of the driving frequency relative to the natural frequency of the oscillator.

The power absorbed is maximum at resonance, when the driving frequency is equal to the natural frequency.

As the driving frequency deviates either way from the natural frequency, the average power absorbed decreases. The difference in values of the driving frequency at which the average power absorbed drops to half its maximum (resonant) value, is called the "band-width" of resonance.

Figure given below shows a graph between the average power absorbed P_{av} and the driving frequency p. The power is maximum for $p=\omega$, where ω is the natural frequency of the oscillator. If ω_1 and ω_2 are the values of p for which P_{av} drops to half its maximum value, then $(\omega_1-\omega_2)$ is the "band-width" of resonance.

Let m be the mass of the oscillator and $F_0 \sin pt$ the driving force. Then, in the steady state, the average absorbed power is given by

$$P_{av}=\frac{mf_0^{\,2}rp^2}{\left(\omega^2-p^2\right)^2+4r^2p^2} \qquad ...(i)$$

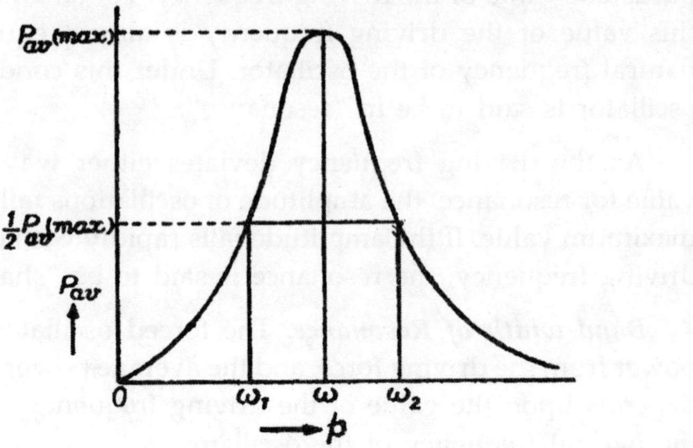

where $f_0 = F_0/m$; $r = b/2m$ (*b* being the damping constant). P_{av} is maximum when $p = \omega$; and the maximum value is

$$P_{av}(\text{max}) = \frac{mf_0^{\,2}}{4r}. \qquad \text{...(ii)}$$

To evaluate the band-width $\omega_1 - \omega_2$, we write

$$P_{av} = \tfrac{1}{2} P_{av}(\text{max})$$

or $$\frac{mf_0^{\,2} rp^2}{(\omega^2 - p^2) + 4r^2p^2} = \frac{1}{2}\;\frac{mf_0^{\,2}}{4r}$$

or $$(\omega^2 - p^2) + 4\,r^2p^2 = 8r^2p^2$$

or $$(\omega^2 - p^2) = 4r^2p^2$$

or $$\omega^2 - p^2 = \pm\, 2\,rp$$

or $$\omega - p = \pm\, \frac{2rp}{\omega + p}$$

If $\omega_2 > \omega_1$, then

$$\omega_2 - p = +\frac{2rp}{\omega + p}$$

and $$\omega_1 - p = -\frac{2rp}{\omega + p}.$$

Subtracting, we get

$$\omega_1 - \omega_2 = \frac{4rp}{\omega + p} = \frac{4r}{\frac{\omega}{p} + 1}$$

If we write $\omega/p \simeq 2r$, then

$$\omega_1 - \omega_2 \simeq 2r$$

or $$\boxed{\omega_2 - \omega_1 = \frac{1}{\tau},}$$

where τ (=1/2r) is the relaxation time. This is the expression for the band-width.

Clearly, smaller the band-width, sharper is the resonance.

Quality Factor of a Driven Oscillator

The quality factor Q of an oscillator is defined by the ratio

$$Q = 2\pi \frac{\text{average energy}}{\text{energy dissipated per period}}.$$

Let us consider a driven harmonic oscillator of mass m subjected to a restoring force —kx, a damping force —$b(dx/dt)$ and a driving force F_0 sin pt. In the steady state, the instantaneous displacement of the oscillator is given by

$$x = \frac{f_0}{\sqrt{\left[(\omega^2 - p^2)^2 + 4r^2p^2\right]}} \sin(pt - \theta).$$

where $f_0 = \frac{F}{m}$, $r = \frac{b}{2m}$, $\omega^2 = \frac{k}{m}$ and tan $\theta \frac{2rp}{\omega^2 - p^2}$. The instantaneous velocity is

$$\frac{dx}{dt} = \frac{f_0 p}{\sqrt{\left[\left(\omega^2 - p^2\right)^2 + 4r^2p^2\right]}} \cos(pt - \theta).$$

Now, the energy stored in the oscillator

=kinetic energy+potential energy

$$= \tfrac{1}{2} m \left(\frac{dx}{dt}\right)^2 + \tfrac{1}{2} kx^2$$

$$= \tfrac{1}{2} m \left(\frac{dx}{dt}\right)^2 + \tfrac{1}{2} m\omega^2 x^2.$$

The rate of dissipation of energy

=damping force × velocity

$$= b\frac{dx}{dt} \times \frac{dx}{dt} = 2mr \left(\frac{dx}{dt}\right)^2$$

$$= \frac{m}{\tau} \left(\frac{dx}{dt}\right)^2,$$

where τ (=1/2r) is the relaxation time.

The oscillator acquires the frequency of the periodic force, so that its time-period $2\pi/p$. Thus,

$$\frac{\text{energy stored}}{\text{period} \times \text{rate of energy dissipation}}$$

$$= \frac{\tfrac{1}{2} m \left(\frac{dx}{dt}\right)^2 + \tfrac{1}{2} m\omega^2 x^2}{\frac{2\pi}{p} \times \frac{m}{\tau} \left(\frac{dx}{dt}\right)^2}$$

$$= \frac{\frac{1}{2} m \frac{f_0{}^2 p^2 \cos^2(pt-\theta)}{(\omega^2-p^2)^2+4r^2p^2} + \frac{1}{2} m\omega^2 \frac{f_0{}^2 p^2 \sin^2(pt-\theta)}{(\omega^2-p^2)^2+4r^2p^2}}{\frac{2\pi}{p} \times \frac{m}{\tau} \frac{f_0{}^2 p^2 \cos^2(pt-\theta)}{(\omega^2-p^2)^2+4r^2p^2}}$$

$$= \frac{\frac{1}{2} p^2 \cos^2(pt-\theta) + \frac{1}{2} p^2 \sin^2(pt-\theta)}{\frac{2\pi}{\tau} p \cos^2(pt-\theta)}.$$

The average values of $\cos^2(pt-\theta)$ as well as of $\sin^2(pt-\theta)$ for one period is $\frac{1}{2}$.

$$\therefore \quad \frac{\text{average energy stored}}{\text{energy dissipated per period}} = \frac{\frac{1}{2}p^2 + \frac{1}{2}\omega^2}{\frac{2\pi p}{\tau}}$$

By definition, the quality factor is therefore

$$Q = 2\pi \times \frac{\frac{1}{2}(p^2+\omega^2)}{\frac{2\pi p}{\tau}}$$

$$= \frac{1}{2}\left(\frac{p^2+p^2}{p^2}\right)(p\tau)$$

$$= \frac{1}{2}[1+(\omega/p)^2](p\tau)$$

This is the exact expression for the Q of a driven harmonic oscillator.

Near resonance, $p = \omega$, and we get

$$Q = \omega\tau.$$

Determination of Frequency of A.C.

The interesting phenomenon of resonance supplies us a method for determining the frequency of A.C. Mains by means of a sonometer. The arrangement consists of a hollow wooden box on which a wire of non-magnetic material is fastened to a peg and passes over a pulley. A weight keeps the wire under tension and presses it against two bridges. A horseshoe permanent magnet is placed over the midpoint of the wire such that the wire lies between the poles of the magnet and is at right angles to the magnetic field. The alternating current from the mains is brought to 6-8 volts by means of a step-down transformer and sent along the wire. A choke-coil is inserted in the circuit which controls the current and prevents the wire from getting heated up.

Principle and Procedure: A current-carrying conductor placed at right angles in a uniform magnetic field is acted upon by a mechanical force which is perpendicular to the directions both of the current and of the field (by Fleming's left-hand rule).

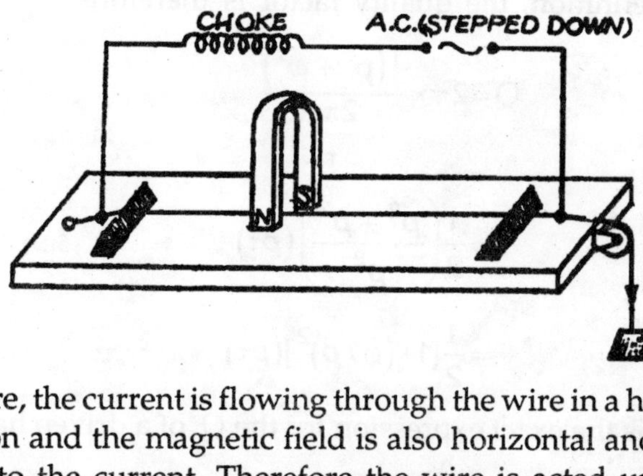

Here, the current is flowing through the wire in a horizontal direction and the magnetic field is also horizontal and at right angles to the current. Therefore the wire is acted upon by a mechanical force in the vertical direction. As the current is alternating, for half the cycle the force is vertically upwards and for next half-cycle vertically downwards.

Thus the wire is acted upon by a periodic force at the frequency, *f*, of the alternating current. Hence it oscillates with the frequency of the current. These are the forced oscillations of the wire.

The wire, however, has a natural frequency n also, which is given by

$$n = \frac{1}{2l}\sqrt{\left(\frac{T}{m}\right)},$$

where l is the length of the wire between the bridges; m its mass per unit length and T the tension applied. If the distance between the two bridges is so adjusted that the natural frequency n of the wire equals the frequency f of the current, resonance will occur and the wire will oscillate with a maximum amplitude. Thus at maximum amplitude, we shall have

$$f = n = \frac{1}{2l}\sqrt{\left(\frac{T}{m}\right)}.$$

To perform the experiment, the distance between the two bridges is adjusted until the amplitude is a maximum. The length l of the wire between the bridges is then measured. The radius of the wire is measured at several places in two mutually perpendicular directions by a screw gauge and the mean radius r is obtained. The value of m is then calculated by the relation

$$m = \pi r^2 d,$$

where d is the density of the material of the wire. The frequency of A.C. mains, f, is then calculated from the last expression.

Transient behaviour in Forced Oscillations

When a damped oscillating system is subjected to an external driving force, it does not reach its final steady state at once, but takes some time. In fact, immediately after the driving force is applied, the system tries to maintain its natural frequency while the driving force tries to impose its own frequency upon the system. Thus there is a tussle between the driving force and the system during which the amplitude of oscillation rises and falls alternately. This transient stage may take quite long if the damping is small. It is known as 'transient effect'.

Let us first consider the (ideal) case when the damping is zero. Let us suppose that the driving force $F_o \cos pt$ is turned on at $t=0$ when the system is at rest. The equation of motion is

$$m\frac{d^2x}{dt^2} + kx = F_o \cos pt$$

or $$\frac{d^2x}{dt^2} + \omega^2 x = f_o \cos pt, \quad \text{...(i)}$$

where $\omega^2 = k/m$ and $f_0 = F_0/m$. Its solution is

$$x = \frac{f_0}{\omega^2 - p^2} \cos pt. \quad \text{...(ii)}$$

If $p < \omega$, the displacement at $t=0$ immediately assumes a positive value. But no system can be displaced by a force in zero time. Again if $P > \omega$, the displacement at $t=0$ immediately assumes a negative value under a positive force, which is absurd. Thus the above solution gives impossible results. The transient, however, explains the facts.

Let x_1 be a solution to eq. (i), and x_2 be a solution to the equation of "free" vibration (in the absence of driving force). Then, we have

$$\frac{d^2x_1}{dt^2} + \omega^2 x_1 = f_0 \cos pt$$

and $$\frac{d^2x_2}{dt^2} + \omega^2 x_2 = 0.$$

Adding these two equations, we get

$$\frac{d^2(x_1 + x_2)}{dt^2} + \omega^2 (x_1 + x_2) = f_0 \cos pt.$$

Thus $x_1 + x_2$, is just as much a solution of the equation of forced motion as is x_1 alone. This means that, at least in the initial state, the motion contains both the free natural frequency ω of the system (corresponding to x_2) and the forced frequency p of the force (corresponding to x_1).

The equation of the free vibration of frequency ω does contain two adjustable constants; an amplitude a and an initial phase ϕ. We can use them to fit conditions at the beginning of the forced motion.

Let us now write the complete solution of the forced-motion eq. (i) as

$$x = a \cos(\omega t+\phi) + b \cos pt. \qquad \text{...(iii)}$$

where $b = \dfrac{f_0}{\omega^2 - p^2}$

At t=0, x=0, so that

$$0 = a \cos \phi + b. \qquad \text{...(iv)}$$

Differentiating eq. (iii), we have

$$\frac{dx}{dt} = -\omega a \sin(\omega t+\phi) - pb \sin pt.$$

At t=0, $\dfrac{dx}{dt} = 0$, so that

$$0 = -\omega a \sin \phi.$$

This means that $\phi = 0$ or π. Hence eq. (iv) gives

$$a = -b \text{ or } b.$$

Putting corresponding values of a and ϕ in eq. (iii), we get

$$x = b(\cos Pt - \cos \omega t). \qquad \text{...(v)}$$

This equation represents the phenomenon of beats. In the complete absence of damping these beats would continue indefinitely and no steady state corresponding to eq. (ii) alone would ever be reached.

The conditions just after $t=0$ can now be explained. If $pt << 1$ and $\omega t << 1$, then

$$\cos pt \approx 1 - \frac{p^2t^2}{2} \text{ and } \cos \omega t \approx 1 - \frac{\omega^2t^2}{2}.$$

Eq. (v) can now be written as

$$x \approx \frac{f_0}{\omega^2 - p^2} \frac{\left(\omega^2 - p^2\right)t^2}{2} = \frac{1}{2} f_0 t^2.$$

This is what we expect. Just after the driving force F_0 cos pt is applied (just before the restoring force comes into play), the mass (system) starts out in the direction of the force with acceleration $f_0 (= F_0/m)$.

In the presence of damping, the complete solution of the forced-motion equation may be written as

$$x = a\, e^{-rt} \cos\left(\omega' t + \phi\right) + A \cos(pt - \theta),$$

where $\omega' = \sqrt{\left(\omega^2 - r^2\right)}$, $A = \dfrac{f_0}{\sqrt{\left[\left(\omega^2 - p^2\right)^2 + 4r^2p^2\right]}}$ and

$\theta = \text{ten}^{-1} \dfrac{2rp}{\omega^2 - p^2}$ This solution represents the transient behaviour of a damped oscillator under a periodic driving force. The motion is like beats, settling down to oscillations of constant amplitude at the driving frequency p (steady state).

The transient behaviour of the oscillator if driven at its own natural frequency ($p = \omega$, resonance). It is a *smooth* growth towards steady amplitude.

8

Forced Electrical Oscillations

Parallel Resonant Circuits

Driven L.C.R Circuit (Forced Electrical Oscillations): An L-C-R circuit fed by an alternating e.m.f. is an example of driven harmonic oscillator.

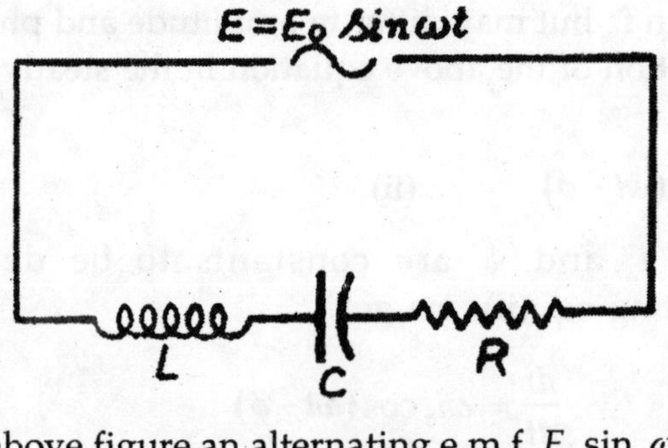

In the above figure an alternating e.m.f. $E_0 \sin \omega t$ has been applied to a circuit containing an inductance L, capacitance C and resistance R in series. Let q be the charge on the capacitor at any instant and i the current in the circuit at that instant. Then the p.d. across the capacitor is q/C, and the back e.m.f. due to self-induction in L is $L(di/dt)$, both being opposite to the applied e.m.f. The effective e.m.f. in the circuit at that instant is, therefore, $E_0 \sin \omega t - \frac{q}{C} - L\frac{di}{dt}$. This, by Ohm's law, must be equal to Ri. That is

$$E_0 \sin \omega t - \frac{q}{C} - L\frac{di}{dt} = Ri$$

or
$$L\frac{di}{dt} + Ri + \frac{q}{C} = E_0 \sin \omega t .$$

Differentiating, we get

$$L\frac{d^2i}{dt^2} + R + \frac{di}{dt} + \frac{1}{C}\frac{dq}{dt} = E_0 \omega \cos \omega t .$$

But $\frac{di}{dt} = i$. Therefore,

$$\boxed{L\frac{d^2i}{dt^2} + R\frac{di}{dt} + \frac{i}{C} = \omega E_0 \cos \omega t.} \quad \text{...(i)}$$

This is the differential equation of the forced electrical oscillations. Once the transient effect dies away and a steady state is set up, the current alternates with the frequency of the applied e. m f., but may differ in amplitude and phase. Hence let the solution of the above equation in the steady state be of the form

$$i = i_0 \sin(\omega t - \phi), \quad \text{...(ii)}$$

where i_o and ϕ are constants to be determined. Differentiating eq. (ii), we get

$$\frac{di}{dt} = \omega i_0 \cos(\omega t - \phi)$$

and
$$\frac{d^2i}{dt^2} = -\omega^2 i_0 \cos(\omega t - \phi) .$$

Substituting for $i, \frac{di}{dt}$ and $\frac{d^2i}{dt^2}$ in eq. (i), we get

$$-L\,\omega^2 i_0 \sin(\omega t-\phi) + R\,\omega i_0 \cos(\omega t-\phi) + \frac{i_0}{C}\text{-}\sin(\omega t-\phi) = \omega E_0 \cos\omega t$$

or $$\left(-L\omega^2+\frac{1}{C}\right) i_0 \sin(\omega t-\phi) + R\,\omega i_0 \cos(\omega t-\phi)$$

$$= \omega E_0 [\cos(\omega t-\phi)+\phi]$$

$$= \omega E_0 [\cos(\omega t-\phi)\cos\phi - \sin(\omega t-\phi)\sin\phi].$$

This equation must be true for all values of t. Hence, equating the coefficients of sin $(\omega t-\phi)$ and cos $(\omega t-\phi)$, we get

$$\left(-L\omega^2+\frac{1}{C}\right) i_0 = -\omega E_0 \sin\phi \qquad \text{...(iii)}$$

and $$R\omega i_0 = \omega E_0 \cos\phi. \qquad \text{...(iv)}$$

Dividing eq. (iii) by eq. (iv), we get

$$\tan\phi = -\frac{-L\omega^2+\dfrac{1}{C}}{R\omega} = \frac{\omega L - \dfrac{1}{\omega C}}{R} \qquad \text{...(v)}$$

Squarring and adding eq. (iii) and (iv), we get

$$i_0^2\left[\left(-L\omega^2+\frac{1}{C}\right)^2 + R^2\omega^2\right] = \omega^2 E_0^2$$

or $$i_0^2\left[R^2+\left(\omega L-\frac{1}{\omega C}\right)^2\right] = E_0^2$$

or $$i_0 = \frac{E_0}{\sqrt{\left[R^2+\left(\omega L-\dfrac{1}{\omega C}\right)^2\right]}}. \qquad \text{...(vi)}$$

Substituting the value of i_0 in eq. (ii), we get

$$i = \frac{E_0}{\sqrt{\left[R^2 + \left(\omega L - \frac{1}{\omega C}\right)^2\right]}} \sin(\omega t - \phi). \qquad \text{...(vii)}$$

This is the equation for the current at any instant in the circuit. The amplitude of the current is

$$i_0 = \frac{E_0}{\sqrt{\left[R^2 + \left(\omega L - \frac{1}{\omega C}\right)^2\right]}}.$$

The quantity $\sqrt{R^2 + \left(\omega L - \frac{1}{\omega C}\right)^2}$ is the 'impedance' Z of the circuit. The quantity $\left(\omega L - \frac{1}{\omega C}\right)$ is the resultant reactance of the circuit which is the difference between the inductive reactance X_L $(= \omega L)$ and the capacitive reactance Xc $(=1/\omega C)$. Thus

$$Z = \sqrt{R^2 + (X_L - X_C)^2}$$

$$i_0 = \frac{E_0}{Z} = \frac{E_0}{\sqrt{R^2 + (X_L - X_C)^2}}. \qquad \text{...(viii)}$$

Eq. (vii) shows that the current i lags in phase from the applied e.m.f. $E_0 \sin \omega t$ by an angle ϕ given by eq. (v), that is,

$$\phi = \tan^{-1} \frac{\omega L - \frac{1}{\omega C}}{R} = \tan^{-1} \frac{X_L - X_C}{R}.$$

The following three cases arise :

(i) When $X_L > X_C$, ϕ is positive, that is, the current lags behind the e.m.f.

(ii) When $X_L<X_C$, ϕ is negative, that is, the current leads the e.m.f.

(iii) When $X_L=X_C$, $\phi=0$, that is, the current is in phase with the e.m.f.

Electrical Resonance: According to eq. (viii), the current have its *maximum* amplitude when

or $X_L - X_C = 0$

or $\omega L - \frac{1}{\omega C} = 0$

or $\omega L = \frac{1}{\omega C}$

or $\boxed{\omega = \frac{1}{\sqrt{LC}}}$.

ω is the (angular) frequency of the applied e.m.f., while $\frac{1}{\sqrt{LC}}$ is the (angular) natural frequency of the circuit in the absence of resistance (damping). Hence we conclude that the maximum amplitude of the current oscillations occurs when the frequency of the applied e.m.f. is exactly equal to the natural {undamped) frequency of the electrical circuit. This is the condition of 'electrical resonance.

At resonance ($X_L=X_C$) the current is always in phase with the applied e.m.f., and its amplitude is determined entirely by the resistance, as follows from eq. (viii).

$i_0 = \frac{E_0}{R}$ (at resonance).

Figure given below shows i_0 as a function of ω for an oscillating *LCR* circuit containing three different values of resistance. It is seen that the smaller the resistance, the sharper the resonance curve.

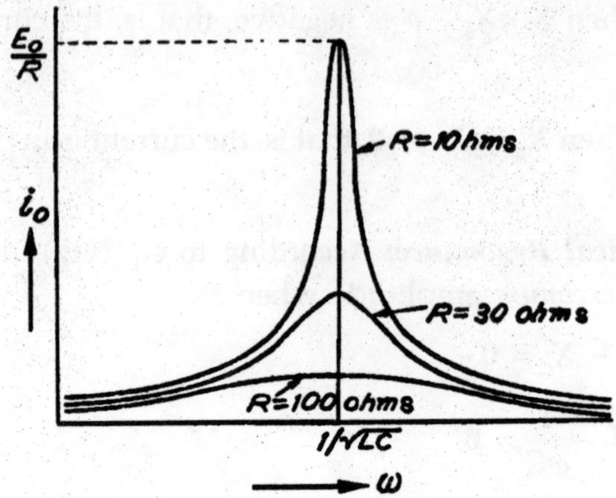

Relaxation Time and Quality Factor of an Oscillating Circuit

The relaxation time τ of an oscillator is an expression of the damping present in the oscillator. It is defined as the time token for the oscillator energy to decay to 1/e of its original value. Higher the damping, smaller the relaxation time.

The quality factor Q of an oscillator is also an expression of damping. It is defined as 2 times the ratio of the energy stored in the oscillator to the energy lost pet period. That is

$$Q = 2\pi \frac{\text{energy stored}}{\text{energy lost per period}}.$$

Evidently, Q is a dimensionless quantity. The higher the damping, the greater is the energy loss per period and smaller is the number Q.

Sharpness of Resonance and Band-width

When an alternating e.m.f. $E_0 \sin \omega t$ is applied to an *L-C-R* circuit, electrical oscillations occur in the circuit with the frequency ω of the applied e.m.f. The amplitude of these oscillations (current amplitude) in the circuit is given by

$$i_0 = \frac{E_0}{\sqrt{R^2 + \left(\omega L - \frac{1}{\omega C}\right)^2}},$$

where $\sqrt{\left[R^2+\left(\omega L-\dfrac{1}{\omega C}\right)^2\right]}$ is the "impedance" Z of the circuit.

At resonance, when the frequency *to* of the applied e.m.f. is equal to the natural frequency $\omega_0\left(=\dfrac{1}{\sqrt{LC}}\right)$ of the circuit, the current amplitude i_0 is a maximum and is equal to *E/R*. Thus at resonance the impedance Z of the circuit is *R*. At other values of ω, the current amplitude i_0 is smaller and the impedance Z is larger than *R*.

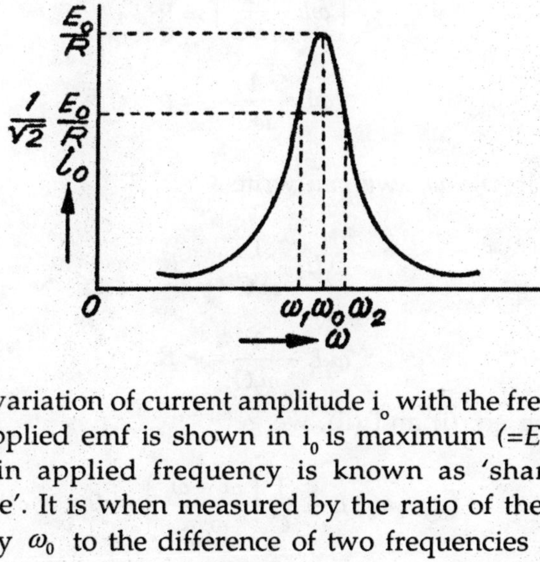

The variation of current amplitude i_0 with the frequency ω of the applied emf is shown in i_0 is maximum ($=E_0/R$) with change in applied frequency is known as 'sharpness of resonance'. It is when measured by the ratio of the resonant frequency ω_0 to the difference of two frequencies ω_1 and ω_2 at which the current falls to $\dfrac{1}{\sqrt{2}}$ of the resonant value. That is,

$$\text{sharpness of resonance} = \frac{\omega_0}{\omega_2-\omega_1}.$$

ω_2 and ω_1 are known as 'half-power frequencies', because at these frequencies the power in the circuit reduces to half its maximum value. The difference of half-power frequencies,

$\omega_2 - \omega_1$, is known as "band-width" of the sharpness of resonance. The smaller is the band-width, the sharper is the resonance.

Expression for the Band-width: At resonant frequency ω_0, the impedance is R. Therefore, at ω_1 and ω_2 it must be $\sqrt{2}R$, that is,

$$Z = \sqrt{R^2 + \left(\omega L - \frac{1}{\omega C}\right)^2} = \sqrt{2}R$$

or
$$R^2 + \left(\omega L - \frac{1}{\omega C}\right)^2 = 2R^2$$

or
$$\left(\omega L - \frac{1}{\omega C}\right)^2 = R^2$$

or
$$\omega L - \frac{1}{\omega C} = \pm R.$$

Thus, if $\omega_2 > \omega_1$, we can write

$$\omega_1 L - \frac{1}{\omega_1 C} = - R \qquad \text{...(i)}$$

and
$$\omega_2 L - \frac{1}{\omega_2 C} = + R. \qquad \text{...(ii)}$$

Adding eq. (i) and (ii), we get

$$(\omega_1 + \omega_2)\, L - \frac{1}{C}\left(\frac{\omega_1 + \omega_2}{\omega_1 \omega_2}\right) = 0$$

or
$$\omega_1 \omega_2 = \frac{1}{LC}.$$

Subtracting eq. (i) from (ii), we get

$$(\omega_1 - \omega_2)\ L + \frac{1}{C} = \left(\frac{\omega_1 - \omega_2}{\omega_1 \omega_2}\right) = 2R$$

or
$$(\omega_1 - \omega_2)\left(L + \frac{\omega_1 - \omega_2}{C\omega_1 \omega_2}\right) = 2R$$

or $$(\omega_1 - \omega_2)\ (L+L) = 2R \qquad \left[\because\ \omega_1\omega_2 = \frac{1}{LC}\right]$$

or $$\boxed{\omega_2 - \omega_1 = \frac{R}{L}} \qquad \text{...(i)}$$

This is the expression for the "band-width".

Now, the quality factor of the *LCR* circuit is given by

$$Q = \frac{1}{R}\sqrt{\frac{L}{C}}$$

$$= \omega_0 \frac{L}{R} \qquad \left[\because\ \omega_0 = \frac{1}{\sqrt{LC}}\right]$$

$$= \frac{\omega_0}{\omega_1 - \omega_2} \qquad \left[\because\ \omega_1 - \omega_2 = \frac{R}{L}\right]$$

or $$\omega_1 - \omega_2 = \frac{\omega_0}{Q}.$$

Thus the band-width is inversely proportional to the *Q* of the circuit. Smaller the band-width (sharper the resonance), higher the quality factor. Thus the quality factor directly measures the sharpness of resonance.

Series and Resonant Circuit

Let an alternating e.m.f. $E = E_0 \sin \omega t$ be applied to a circuit having an inductance *L*, capacitance *C* and resistance *R* all joined in series.

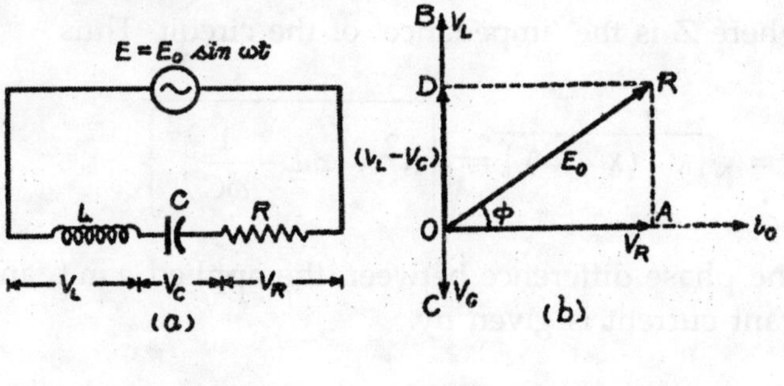

The instantaneous current is the same in *L*, *C* and *R*, and the *vector* sum of the potential differences across the three components is equal to the applied e.m.f.

The peak value (amplitude) of the current in the circuit can be obtained by a vector method. Let *OA*, *OB* and *OC* represent the magnitude and phase of the peak potential differences V_R, V_L and V_c across *R*, *L* and *C* respectively. Their magnitudes are

$$V_R = R\, i_0,\ V_L = X_L\, i_0 = \omega L i_0 \text{ and } V_C = X_C\, i_0 = i_0/\omega C,$$

where i_0 is the peak value of the current and X_L and X_C are the inductive reactance and the capacitive reactance respectively. The magnitude of the resultant of $\vec{V}_L$ and $\vec{V}_C$, which are opposite in phase, is equal to their difference and can be represented by *OD*, where *OD*=*OB*–*OC*. The diagonal *OR* of the rectangle *OARD* then represents, in magnitude and direction, the resultant of V_R, V_L and V_C and hence E_0. Now

$$OR^2 = OA^2 + OD^2$$

$$\therefore\ E_0^2 = V_R^2 + (V_L - V_C)^2 = (R\, i_0)^2 + (X_L\, i_0 - X_C\, i_0)^2$$

$$\text{or}\ \ i_0^2 = \frac{E_0^2}{R^2 + (X_L - X_C)^2}$$

$$\text{or}\ \ i_0^2 = \frac{E_0^2}{\sqrt{[R^2 + (X_L - X_C)^2]}} = \frac{E_0}{Z} \qquad \text{...(i)}$$

where Z is the 'impedance' of the circuit. Thus

$$Z = \sqrt{\{R^2 + (X_L - X_C)^2\}} = \sqrt{\left\{R^2 + \left(\omega L - \frac{1}{\omega C}\right)^2\right\}} \qquad \text{...(ii)}$$

The phase difference between the applied e.m.f. and the resultant current is given by.

$$\tan\phi = \frac{V_L - V_C}{V_R} = \frac{V_L - V_C}{R} = \frac{\omega L - \dfrac{1}{\omega C}}{R}. \qquad \text{...(iii)}$$

The eq. (ii) shows that the impedance Z of the circuit is a minimum when

$$X_L = X_C$$

and is equal to the resistance *R* in the circuit. Eq. (i) shows that under this condition the series circuit allows the maximum peak current *(*$=E_0/R$*)* to flow through it, and this current is in phase ($\phi=0$ as shown by eq. iii) with the applied e.m.f. The circuit is then called the 'series resonant circuit.'

The frequency (of the applied e.m.f.) at which the resonance takes place is given by

$$X_L = X_C$$

or
$$\omega L = \frac{1}{\omega C}$$

or
$$\omega = \frac{1}{\sqrt{LC}},$$

which is the natural frequency of the circuit. Thus the resonant frequency depends on the *product of Land C,* and does not depend on R.

Acceptor Circuit: The series resonant circuit is often called an 'acceptor' circuit. This is because the impedance of the circuit is minimum at resonance so that it most readily accepts that current out of many currents whose frequency is equal to its resonant (natural) frequency. In radio receivers, the resonant frequency of the circuit is tuned (by changing C or I,) to the, frequency of the signal desired to be detected.

Voltage Magnification: At resonance, the (peak) current through the circuit is

$$i_0 = \frac{E_0}{R},$$

so that the (peak) voltage across the inductance is

$$V_L = X_L\, i_0 = \omega L \times \frac{E_0}{R}$$

$$= \frac{\omega L}{R}(E_0)$$

$$= QE_0,$$

where Q is the 'quality factor' of the circuit. Thus at resonance the voltage drop across L (which is also the voltage across C) is Q times the applied voltage. Hence the chief characteristic of the series resonant circuit is 'voltage magnification'.

Parallel Resonant Circuit: Let an alternating e.m.f. $E = E_0 \sin \omega t$ be applied to a circuit having an inductance L and resistance R in parallel with a capacitance C. The peak value of the current in the inductance (and resistance) is given by

$$i_1 = \frac{E_0}{\sqrt{(R^2 + X_L{}^2)}}, \text{ (where } X_L = \omega_L\text{)}$$

and lags behind E_0 by an angle ϕ, where $\tan \phi = \frac{X_L}{R} = \frac{\omega L}{R}$. The peak current in the capacitance is given by

$$i_2 = \frac{E_0}{X_C}, \left(\text{where } X_C = \frac{1}{\omega C}\right)$$

and leads E_0 by 90°.

In figure given below, i_1 and i_2 have been represented by the vectors OA and OB. The resultant peak current i_0 is given by the vector sum of i_1 and i_2 and is represented by the diagonal OR of the parallelogram $OARB$. The current i_0 is known as the 'supply current' of the circuit.

The impedance of the parallel circuit will then be given by

$$Z = \frac{E_0}{i_0}.$$

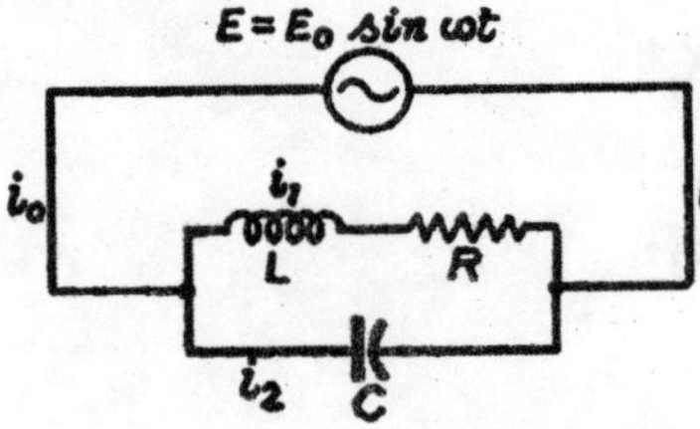

(a)

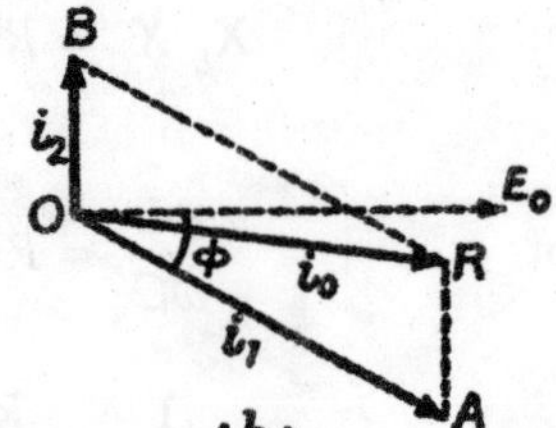

(b)

Now, the parallel circuit is said to be in resonance when the supply current i_0 is in phase with the applied e.m.f. E_0. The circuit then acts like a pure resistance.

At resonance the diagonal OR will lie along the direction of E_0. Now the component of i_1 along BO is $i_1 \sin\phi$, and along OR is $i_1 \cos\phi$. Since the vector sum of i_1 and i_2 is along OR, it follows that

$$i_2 = i_1 \sin\phi \qquad \text{...(i)}$$

and

$$i_0 = i_1 \cos\phi . \qquad \text{...(ii)}$$

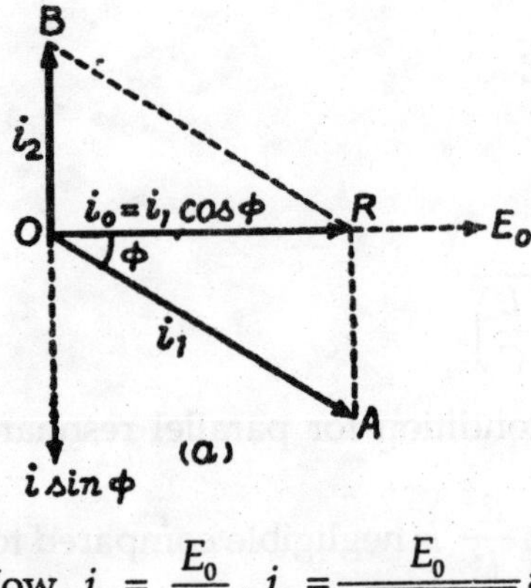

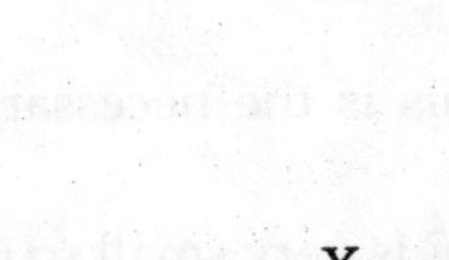

(b)

Now, $i_2 = \dfrac{E_0}{X_C}$, $i_1 = \dfrac{E_0}{\sqrt{R^2 + X_L{}^2}}$ and $\sin\phi = \dfrac{X_L}{\sqrt{R^2 + X_L{}^2}}$.

Substituting these values in eq. (i), we get

$$\frac{E_0}{X_C} = \frac{E_0}{\sqrt{(R^2 + X_L{}^2)}} = \frac{X_L}{\sqrt{(R^2 + X_L{}^2)}}$$

or $$X_L X_C = R^2 + X_L^2. \quad \text{...(iii)}$$

or $$\frac{\omega L}{\omega C} = R^2 + \omega^2 L^2$$

or $$\frac{1}{LC} = \frac{R^2}{L^2} + \omega^2$$

or $$\omega^2 = \frac{1}{LC} - \frac{R^2}{L^2}$$

or $$\omega = \sqrt{\left(\frac{1}{LC} - \frac{R^2}{L^2}\right)}. \quad \text{...(iv)}$$

This is the resonant frequency of the circuit. Obviously, for parallel resonance to occur, $\sqrt{\left(\frac{1}{LC} - \frac{R^2}{L^2}\right)}$ should be real, *i.e.*

$$\frac{R^2}{L^2} < \frac{1}{LC}$$

or $$R^2 < \frac{L}{C}$$

or $$R < \sqrt{\left(\frac{L}{C}\right)}.$$

This is the necessary condition for parallel resonance.

If R is very small so that $\frac{R^2}{L^2}$ *is* negligible compared to $\frac{1}{LC}$, the eq. (iv) for resonant frequency becomes

$$\omega = \frac{1}{\sqrt{LC}},$$

which is same as for series resonance.

The impedance of the circuit is

$$Z = \frac{E_0}{i_0}.$$

At resonance, we have from eq. (ii)

$$i_0 = i_1 \cos\phi.$$

But $i_1 = \dfrac{E_0}{\sqrt{(R^2 + X_L^2)}}$ and $\cos\phi = \dfrac{R}{\sqrt{(R^2 + X_L^2)}}$.

$$\therefore \quad i_0 = \frac{E_0}{\sqrt{(R^2 + X_L^2)}} \frac{R}{\sqrt{(R^2 + X_L^2)}} = \frac{E_0 R}{R^2 + X_L^2}.$$

But $R^2 + X_L^2 = X_L X_C$ by eq. (iii).

$$\therefore \quad i_0 = \frac{E_0 R}{X_L + X_C} = \frac{E_0 R}{\omega L / \omega C} = \frac{E_0 RC}{L}.$$

This expression shows a very interesting feature. The current i_0 is proportional to R (not to $1/R$). Since R is very small, the current i_0 is also very small at resonance. If R is negligible, the current is zero.

The impedance (or dynamic resistance) at resonance is

$$Z = \frac{E_0}{i_0} = \frac{L}{CR}$$

Thus smaller the resistance R, larger is the impedance. If R is negligible, the impedance is infinite at resonance.

Rejector Circuit: The parallel resonant circuit does not allow the current of the same frequency as the natural frequency of the circuit. Thus it can be used to suppress the current of this particular frequency out of currents of many other frequencies. Hence the circuit is known as a 'rejector' or 'filter' circuit. (In contrast, the series circuit has a minimum impedance or maximum current at resonance, and is, therefore, called as an 'acceptor circuit').

Current Magnification: The (peak) current at resonance from the supply source is

$$i_0 = \frac{E_0 RC}{L}.$$

The current circulating in the circuit, say in the capacitor branch is

$$i_2 = E_0/X_C = E_0 \omega C.$$

Thus $$\frac{i_2}{i_0} = \frac{E_0 \omega C}{E_0 RC/L} = \frac{\omega L}{R}.$$

Thus the parallel resonant circuit at resonance shows current magnification of $\omega L/R$, which is the same as the voltage magnification shown by series resonant circuit.

The parallel resonant circuit is widely used in radio due to its high impedance at resonance and the current magnification which it can give.

Mechanical and Electrical Impedance

The mechanical impedance of a physical system subjected to a driving force is defined as the ratio of the driving force to the associated velocity of displacement.

Thus, if F is the driving force and v the velocity, then the mechanical impedance is

$$Z_m = \frac{F}{v}.$$

The concept of mechanical impedance is similar to the concept of electrical resistance which is the ratio of an applied e.m.f. to the associated current (rate of flow of charge), Ohm's law is ordinarily a scalar relation ($R=E/i$) in which the e.m.f. and the current are always in phase. However, the presence of inductance *L* or capacitance C, or both, alongwith the resistance *K* in the circuit introduces a phase difference between the applied e.m.f. and the current. Ohm's law then takes the vector form

$$Z_e = \frac{E}{i},$$

where Z_e is the (electrical) impedance of the circuit. It is the vector sum of the effective resistances of *R*, *L* and *C* in the circuit. Z is a complex quantity.

Impedance of an Electrical Circuit: Suppose an alternating e.m.f. E of frequency — is applied across a resistance *R*, inductance *L* and capacitance *C* in series. The balance of e.m.f.'s is given by

$$Ri + L\frac{di}{dt} + \frac{q}{C} = E. \qquad \text{...(i)}$$

and the current through the circuit is given by

$$i = i_0 e^{j\omega t}$$

The voltage and current across the resistance are in phase.

The voltage across the inductance is

$$L\frac{di}{dt} = L\frac{d}{dt}\left(i_0 e^{j\omega t}\right)$$

$$= j\omega L\, i_0 e^{j\omega t}$$

$$= j\omega L i.$$

ωL is the effective resistance (reactance) presented by *L* to the current of frequency ω. The presence of *j* indicates that the voltage w*Li* across *L* is 90° ahead the current through the circuit.

The voltage across the capacitor is

$$\frac{q}{C} = \frac{1}{C}\int i\,dt$$

$$= \frac{1}{C} i_0 \int e^{j\omega t}\,dt$$

$$= \frac{1}{j\omega C} i_0\, e^{j\omega t}$$

$$= \frac{1}{j\omega C}\, i$$

$$= -\frac{j}{\omega C}\, i. \qquad \left[\because \frac{1}{j} = -j\right]$$

Again, $\frac{1}{\omega C}$ is the effective resistance (reactance) presented by C to the current of frequency ω. The presence of $-j$ indicates that the voltage $\frac{i}{\omega C}$ across the capacitor lags the current by 90°.

Hence eq. (i) can be written as

$$Ri + j\omega L\, i - \frac{j}{\omega C} i = E$$

or
$$\frac{E}{i} = R + j\left(\omega L - \frac{1}{\omega C}\right).$$

But $\frac{E}{i}$ is the (complex) impedance of the circuit. Therefore,

$$Z_e = R + j\left(\omega L - \frac{1}{\omega C}\right).$$

The. magnitude of the impedance is

$$Z_e = \sqrt{R^2 + \left(\omega L - \frac{1}{\omega C}\right)^2}.$$

Its phase is given by

$$\tan \phi = \frac{\omega L - \dfrac{1}{\omega C}}{R}.$$

ϕ is the phase difference between the applied e.m.f. and the current in the circuit.

Impedance of a Mechanical Circuit: By analogy, we can write the impedance of a mechanical circuit having mass m, spring factor k and damping factor b under an applied force F of frequency p. We have

$$L \equiv m$$

$$C \equiv \frac{1}{k}$$

$$R \equiv b$$

$$\omega \equiv p.$$

The mechanical impedance is

$$Z_m = b + j\left(pm - \frac{k}{p}\right).$$

For resonance (minimum impedance), we have

$$pm - \frac{k}{p} = 0$$

or $$p^2 = \frac{k}{m}$$

or $$p = \sqrt{\left(\frac{k}{m}\right)}$$

The frequency of the mechanical system is

$$n = \frac{p}{2\pi} = \frac{1}{2\pi}\sqrt{\left(\frac{k}{m}\right)}$$

Problems

1. A circuit has an inductance of $\frac{1}{\pi}$ *henry and resistance 100* Ω. *An A.C. supply of 50 cycles is applied to it. Calculate the reactance and the impedance offered by the circuit.*

Solution: The (inductive) reactance is

$X_L = \omega L$.

Here $\omega = 2\pi n = 2\pi \times 50 = 100\pi$ /rad/sec and $L = \frac{1}{\pi}$ henry.

$$\therefore \quad X_L = 100\pi \times \frac{1}{\pi} = 100 \text{ ohm.}$$

The impedance is

$$Z = \sqrt{(R^2 + X_L^2)}$$

$$\sqrt{[(100)^2 + (100)^2]} = 141.4 \text{ ohm.}$$

2. A coil having an inductance of 0.14 henry and a resistance of 12 ohm is connected across a 110-volt, 25-cycle line. Compute the impedance of the coil, current in the coil and the phase angle between the current and the supply voltage.

Solution. The (inductive) reactance is

$$X_L = \omega L = 2\pi n L = 2 \times \tfrac{22}{7} \times 25 \times 0.14 = 22 \text{ ohm.}$$

The impedance is

$$Z = \sqrt{(R^2 + X_L^2)} = \sqrt{\left[(12)^2 + (22)^2\right]} = 25.1 \text{ ohm.}$$

The current is

$$i = \frac{E}{Z} = \frac{110 \text{ volt}}{25.1 \text{ ohm}} = 4.38 \text{ amp.}$$

The phase angle is

$$\phi = \tan^{-1}\frac{X_L}{R} = \tan^{-1}\left(\frac{22}{12}\right)$$

$$= \tan^{-1}(1.83) = 61.3^\circ.$$

3. *A series LCR circuit has L= 1 mH, C=0.1 μF and R=10 ohm. Calculate the resonant frequency of the circuit. What is the separation between the two half-power points and hence the sharpness?*

Solution: The resonant angular frequency of the circuit is given by

$$\omega_0 = \frac{1}{\sqrt{(LC)}}.$$

Here L=1 mH=10^{-3} H and C=0.1 μF=10^{-7} F.

$$\therefore \quad \omega_0 = \frac{1}{\sqrt{(10^{-3}\times 10^{-7})}} = 10^5 \text{ rad/sec.}$$

The Separation between the tow half-power points (band-width) the given by

$$\omega_2 - \omega_1 = \frac{R}{L}$$

$$= \frac{10 \text{ ohm}}{10^{-3} \text{ henry}} = 10^4 \text{ rad/sec.}$$

The sharpness is

$$\frac{\omega_0}{\omega_2 - \omega_1} = \frac{10^5}{10^4} = 10.$$

9

Damped Harmonic Oscillations

Whenever a system is set into oscillations under no external influence, its amplitude of oscillation gradually decreases to zero. This is because the oscillating system is always subject to frictional forces arising from air resistance or from within the system itself. The work done against these forces is dissipated out of the system as heat. Therefore the mechanical energy of the system decreases with time, and eventually the system comes to rest.

As an example, if we displace a pendulum from its equilibrium position, it will oscillate with a decreasing amplitude and finally come to rest in the equilibrium position.

In most cases, the energy lost from the oscillating system in each cycle is proportional to the total energy of the system. Then each cycle is similar to the previous cycle but of smaller amplitude. This happens when the frictional force is proportional to the velocity of the system.

Equation for Damped Harmonic Oscillator: Let us consider a body of mass m oscillating under a spring-force of constant k. Let x be the displacement of the body from the equilibrium position at any instant. The instantaneous velocity is dx/dt. The forces acting on the body at this instant are:

(i) A restoring force proportional to the displacement, but acting in the opposite direction, which may be written as

$$-k.x,$$

where k is force-constant.

(ii) A frictional (or damping) force proportional to the opposite to the direction of motion, which may be written as

$$-b\frac{dx}{dt}$$

where b is a positive constant.

The total instantaneous force acting on the body is therefore

$$F = -kx - b\frac{dx}{dt}$$

But by Newton's law, $F=m\,(d^2x/dt^2)$, where m is the mass of the body and d^2x/dt^2 is the instantaneous acceleration. Then

$$m\frac{d^2x}{dt^2} = -kx - b\frac{dx}{dt}$$

or

$$\frac{d^2x}{dt^2} + \frac{b}{m} + \frac{dx}{dt} + \frac{k}{m}x = 0.$$

Putting $\frac{b}{m} = 2r$ and $\frac{k}{m} = \omega^2$, we get

$$\boxed{\frac{d^2x}{dt^2} + 2r\frac{dx}{dt} + \omega^2 x = 0.} \quad \text{...(i)}$$

This is a homogeneous linear differential equation of the second order.

In order to solve this equation, we try the following solution:

$$x = Ce^{\alpha t} \quad \text{..(ii)}$$

where C and a are arbitrary constants. Differentiating eq. (ii) with respect *to t*, we get

$$\frac{dx}{dt} = C\alpha e^{\alpha t}$$

and
$$\frac{d^2x}{dt^2} = C\alpha^2 e^{\alpha t}.$$

Substituting the values of $\frac{d^2x}{dt^2}$, $\frac{dx}{dt}$ and x in eq. (i), we get

$$C\alpha^2 e^{\alpha t} + 2rC\alpha e^{\alpha t} + \omega^2 C e^{\alpha t} = 0$$

or
$$Ce^{\alpha t}\left(\alpha^2 + 2r\alpha + \omega^2\right) = 0$$

or
$$\alpha^2 + 2r\alpha + \omega^2 = 0$$

or
$$\alpha = -r \pm \sqrt{\left(r^2 - \omega^2\right)}.$$

The general solution of eq. (1) is obtained by putting these two values of x eq. (ii). Thus

$$x = C_1 e^{\left\{-r+\sqrt{\left(r^2-\omega^2\right)}\right\}t} C_2 e^{\left\{-r-\sqrt{\left(r^2-\omega^2\right)}\right\}t}, \qquad \text{...(iii)}$$

where C_1, and C_2, ire arbitrary constants. The actual form of this 'equation depends open whither r^2 is greater than, equal to, or less than, ω^2 Thus are three cases :

First Case—Heavy Damping ($r^2 > \omega^2$): In this case $\sqrt{\left(r^2 - \omega^2\right)}$ is real and less than r. Therefore in eq. (iii), both the exponents $\left\{-r+\sqrt{\left(r^2-\omega^2\right)}\right\}$ and $\left\{-r-\sqrt{\left(r^2-\omega^2\right)}\right\}$ are negative. It means that the displacement x of the particle continuously decreases with time. That is, the particle when once displaced returns to its equilibrium position quite slowly without performing any oscillation. Such a motion is called "overdamped" or 'aperiodic' motion. It is shown by a pendulum moving in thick oil or by a moving-coil galvanometer shunted by a very low resistance.

Second Case—Critical Damping $(r^2 = \omega^2)$: If we put $r^2 = \omega^2$ in eq. (iii), then this solution does not satisfy eq. (i). Hence we consider the case when $\sqrt{(r^2 - \omega^2)}$ is not zero but is a very small quantity h. The eq. (iii) then gives

$$
\begin{aligned}
x &= C_1 e^{(-r+h)t} C_2 e^{(-r-h)t} \\
&= e^{-rt}(C_1 e^{ht} + C_2 e^{-ht}) \\
&= e^{-rt}\{C_1(1+ht+\ldots)+C_2(1-ht+\ldots)\} \\
&= e^{-rt}\{(C_1 + C_2)+ht\ (C_1 - C_2)\} \\
&= e^{-rt}(A+Bt),
\end{aligned}
$$

where $A=(C_1+ C_2)$ and $B=h\ (C_1- C_2)$.

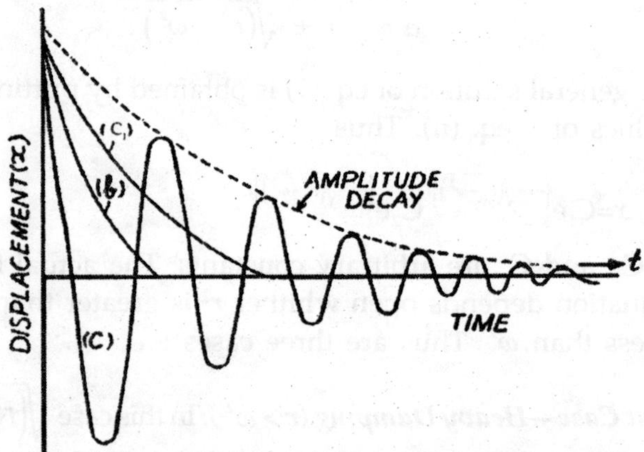

The graph of this equation is the curve (b) in figure above. It is clear that in this case the particle tends to acquire its position of equilibrium much more rapidly than when $r^2 > \omega^2$ (case I). Such a motion is called "critically damped" motion. It is often desirable to arrange for critical damping in many pointer instruments so that the pointer moves to the correct position and returns to zero in minimum time.

Third Case : Light Damping **($r^2 > \omega^2$):** This is the actual case of a damped harmonic oscillator.

In this case $\sqrt{(r^2 - \omega^2)}$ is imaginary. Let us write

$$\sqrt{(r^2 - \omega^2)} = j\sqrt{(\omega^2 - r^2)} = j\omega,'$$

where $\omega' = \sqrt{(\omega^2 - r^2)}$ and $j = \sqrt{(-1)}$. The eq.(iii) now becomes

$$x = C_1 e^{(-r+j\omega')t} + C_2 e^{(-r-j\omega')t}$$

$$= e^{-rt}\left(C_1 e^{j\omega' t} + C_2 e^{-j\omega' t}\right)$$

$$= e^{-rt}\{C_1(\cos\omega' t + j\sin\omega' t) + C_2(\cos\omega' t - j\sin\omega' t)\}$$

$$= e^{-rt}\{(C_1 + C_2)\cos\omega' t + j(C_1 - C_2)\sin\omega' t\}.$$

Putting $C_1 + C_2 = a\sin\phi$ and $j(C_1 - C_2) = a\cos\phi$, where a and ϕ are arbitrary constants, the last expression gives

$$x = e^{-rt}\,(a\sin\phi\cos\omega' t + a\cos\phi\sin\omega' t)$$

or

$$\boxed{x = ae^{-rt}\sin(\omega' t + \phi).} \quad \text{...(iv)}$$

This is the solution of the differential equation of the damped harmonic oscillator. From this we can find amplitude, frequency, energy, etc.

Amplitude: The amplitude of the oscillations is ae^{-rt}, which is *not* constant, but instead decays exponentially with time due to the damping factor e^{-rt}. This is an 'under-damped' motion and is represented by curve (c) in Figure given above.

The time-interval in which the amplitude of oscillation falls to $1/e$ of its initial value is called the 'mean lifetime' of the oscillator. The mean lifetime comes out to be $1/r$ (or $2m/b$). If damping were absent ($r=0$), the mean lifetime would be infinite *i.e.* the amplitude would have remained constant.

Time-Period: The time-period of damped oscillation is

$$T=\frac{2\pi}{\omega'}=\frac{2\pi}{\sqrt{(\omega^2-r^2)}},$$

which is slightly greater than $2\pi/\omega$ (the period in the absence of damping). Thus due to damping, the period is slightly increased, or the frequency is slightly decreased. The damped frequency is

$$\frac{1}{2\pi}\sqrt{(\omega^2-r^2)},$$

where $\omega=\sqrt{\left(\frac{k}{m}\right)}$ and $r=\frac{b}{2m}$

Logarithmic Decrement

This measures the rate at which the amplitude dies away. Suppose in the expression

$x=ae^{-rt}\sin(\omega' t+\phi)$

we choose $\phi=\frac{\pi}{2}$, and we write

$x=a_0\, ae^{-rt}\cos\omega' t,$

so that $x=a_0$ at $t=0$. Let a_1, a_2, a_3 ..... be the amplitudes at time $t=T$, $2T$. $3T$,... where T is the period of oscillation. Then

$$x_1=a_0\, e^{-rT}$$

$$a_1=a_0\, e^{-2rT}$$

$$a_1=a_0\, e^{-3rT}$$

From this, we see that

$$\frac{a_0}{a_1}=\frac{a_1}{a_2}=\frac{a_2}{a_3}=\ldots e^{rT}=e^{\lambda}$$

where λ $(=rT=bT/2m)$ is the "logarithmic decrement . Clearly,

$$\lambda = \log_e \frac{a_0}{a_1} = \log_e \frac{a_1}{a_2} = \log_e \frac{a_2}{a_3} = ...$$

Thus logarithmic decrement is the logarithm of the ratio of two amplitudes of oscillation which are separated by one period.

Power Dissipation in Damped Harmonic Oscillator

Whenever a system is set into oscillations, it is subject to frictional (damping) forces arising from air resistance or from within system itself. These forces oppose the motion of the system. The work done against these forces is dissipated out of the system as heat. Therefore, the mechanical energy of the system continuously decreases with time, and the amplitude of oscillation gradually decays to zero. Let us obtain an expression for this power dissipation from the oscillator.

At any instant t, the displacement of a damped harmonic oscillator is given by

$$x = ae^{-rt} \sin(\omega' t + \phi), \qquad ...(i)$$

where a and ϕ are arbitrary constants, r is a damping constant and $\omega' = \sqrt{(\omega^2 - r^2)}$ is the angular frequency of the (damped) oscillator. $\omega\left[= \sqrt{(k/m)}\right]$ is the angular frequency of undamped oscillator, k being the force-constant.

The instantaneous velocity of the oscillator is

$$u = \frac{dx}{dt} = ae^{-rt}[-r \sin(\omega' t + \phi) + \omega' \cos(\omega' t + \phi)].$$

In practice, the damping is very small so that $r << \omega$. Then the term $-r \sin(\omega' t + \phi)$ in the equation for u can be neglected and we can write

$$u = ae^{-rt}\, \omega' \cos(\omega' t + \phi). \qquad ...(ii)$$

Now, the total mechanical energy of the oscillator may be written as

$$E = \text{kinetic energy} + \text{potential energy}$$

$$= \tfrac{1}{2}\, mu^2 + \tfrac{1}{2}\, kx^2.$$

Substituting the values of x and u from eq. (i) and (ii), we get

$$E = \tfrac{1}{2}\, ma^2 e^{-2rt}\, \omega'^2 \cos^2(\omega' t + \phi) + \tfrac{1}{2}\, ka^2 e^{-2rt} \sin^2(\omega' t + \phi).$$

Again, since $r << \omega$, we can put

$$\omega'^2 = \omega^2 - r^2 = \omega^2 = \frac{k}{m}$$

Then we have

$$E = \tfrac{1}{2}\, ma^2 e^{-2rt}\left(\frac{k}{m}\right)\cos^2(\omega' t + \phi) + \tfrac{1}{2}\, ka^2 e^{-2rt} \sin^2(\omega' t + \phi)$$

$$= \tfrac{1}{2}\, ka^2 e^{-2rt}\, [\cos^2(\omega' t + \phi) + \sin^2(\omega' t + \phi)]$$

$$= \tfrac{1}{2}\, ka^2 e^{-2rt}. \qquad \text{...(iii)}$$

This shows that the energy of the oscillator decreases with time.

Now, the rate at which the energy is lost is the 'power dissipation' P. Thus

$$P = -\frac{dE}{dt} = -\tfrac{1}{2}\, ka^2 e^{-2rt}\,(-2r)$$

$$= 2r\,(\tfrac{1}{2}\, ka^2 e^{-2rt}).$$

This, using eq. (iii) becomes

$$\boxed{P = 2rE.} \qquad \text{..(iv)}$$

Relaxation Time

Let us now introduce 'relaxation time'. It is the time taken for the total mechanical energy to decay to l/e of its original value. If the energy is E_o at t=0, then eq. (iii) gives

$$E_0 = \tfrac{1}{2}\, ka^2.$$

Eq. (iii) now becomes

$$E=E_0e^{-2rt}. \qquad ...(iv)$$

Now, if τ be the relaxation time, then at t= τ . we shall have

$E=\frac{E_0}{e}$. Then, from the last equation, we have

$$\frac{E_0}{e} = E_0\, e^{-2r\tau}$$

or $$e^{-1} = E_0\, e^{-2r\tau}$$

or $$-1 = -2r\tau$$

or $$\boxed{\tau = \frac{1}{2r}} \qquad ...(vi)$$

Making this substitution in eq. (iv) and (v), we get

$$\boxed{P = \frac{E}{\tau}} \qquad ..(vii)$$

and $$\boxed{E = E_0^{e-t/\tau}} \qquad ...(viii)$$

The dissipated energy appears as heat in the oscillating system.

Quality Factor

The quality factor Q of an oscillating system is a measure of damping, or the rate of energy decay, of the system. It is defined as 2π times the ratio of the energy stored in the system to the energy lost per period. That is

$$Q=2\pi\frac{\text{energy stored in system}}{\text{energy loss per period}}$$

Evidently, Q is a dimensionless quantity. Higher the damping, greater is the energy loss per period and smaller is the number Q.

Mean Lifetime

The mean lifetime of a damped oscillator is the time when for the amplitude of oscillation ro decay to 1/e of the initial value.

Relation between Quality Factor and Relaxation Time

The quality factor is defined by

$$Q = 2\pi \frac{\text{energy stored in the system}}{\text{energy loss per period}}$$

If E is the total mechanical energy of the oscillator and P is the rate of energy decay (power dissipation): then we have

$$Q = 2\pi \frac{E}{PT}.$$

where T is the period. Now, if τ be the relaxation time, then

$$P = \frac{E}{\tau}$$

so that
$$Q = 2\pi \frac{E}{(E/\tau)T}$$

But $\frac{2\pi}{T} = \omega$, the angular frequency of the oscillator. Therefore

$$\boxed{Q = \omega\tau.}$$

This is the relation between Q and the relaxation time. For an undamped oscillator, τ is infinite, and so Q is infinite.

The Q of an excited atom is about 10^8, while for a violin string it is only about 10^3 due to a much larger damping in the string.

Difference between Free, Damped and Angular Oscillation

Let m be the mass of the oscillating system, k the force constant (spring-factor) and b the damping constant.

Oscillation	Equation of Motion	Solution	Angular Frequency	Energy
Free	$\frac{d^2x}{dt^2}+\omega^2 x=0$	$x = a \sin(\omega t + \phi)$	$\sqrt{\left(\frac{k}{m}\right)}$	$\frac{1}{2}ka^2$
	where $\omega^2 = k/m$			
Damped	$\frac{d^2x}{dt^2} + 2r\frac{dx}{dt}+\omega^2 x=0$	$x = ae^{-rt}\sin(\omega' t+\phi)$	$\sqrt{\left(\frac{k}{m}-\frac{b^2}{4m^2}\right)}$	$\frac{1}{2}ka^2e^{-2rt}$
	where r = $b/2m$	where $\omega' = \sqrt{(\omega^2 - r^2)}$		
Angular	$\frac{d^2\theta}{dt^2}+\omega^2\theta = 0$	$\theta = \theta_0 \sin(\omega' t+\phi)$	$\sqrt{\left(\frac{c}{I}\right)}$	$\frac{1}{2}I\omega^2\theta_0^2$
	where $\omega^2 = c/I$ $= \frac{\text{torsional constat}}{\text{moment of inertia}}$			

Problems

1. The differential equation of an oscillating system is

$$\frac{d^2x}{dt^2} + 2r\frac{dx}{dt} + \omega^2 x = 0.$$

If $\omega >> r$, then calculate the time in which (i) amplitude becomes 1/e of its initial value, (ii) energy becomes 1/e of the initial value,(iii) Energy becomes $1/e^4$ of its initial value.

Solution: The given differential equation with the condition $\omega >> r$, is the equation of motion of an under-damped harmonic oscillator whose solution is

$$x \Rightarrow ae^{-rt} \sin(\omega' t + \phi)$$

where a and ϕ are arbitrary constants and $\omega' = \sqrt{(\omega^2 - r^2)}$. Thus the amplitude of oscillation is ae^{-rt}.

(i) The damped amplitude at any instant is ae^{-rt}. Let a_0 be the amplitude at t=0 and a_0/e at t. Then we have

$$a_0 = a$$

and $$\frac{a_0}{e} = ae^{-rt} = a_0 e^{-rt}.$$

From this, we have

$$e^{-1} = e^{-r1}$$

or $$-1 = -rt$$

or $$t = \frac{1}{r} \text{ sec.} \qquad \text{(mean life time)}$$

(ii) The instantaneous energy of the damped oscillations is given by

$$E = E_0 e^{-2rt}$$

When the energy falls to E_0/e, we have

or $$\frac{E_0}{e} = E_0 e^{-2rt}$$

or $$e^{-1} = e^{-2rt}$$

or $$t = \frac{1}{2r} \text{ sec.} \qquad \text{(relaxation time)}$$

(iii) When the energy falls to E_0/e^4, we have

or $$\frac{E_0}{e^4} = E_0 e^{-2rt}$$

or $$e^{-1} = e^{-2rt}$$

or $$t = \frac{1}{2r} \text{ sec.}$$

2. *The equation of motion of a system is given by*

$$\frac{d^2y}{dt^2} + \frac{2}{\tau}\frac{dy}{dt} + \omega^2 y = 0.$$

Explain the significance of the various terms of the equation and show that π *has the dimensions of time. Solve the equation and discuss the motion of the system when* $\omega^{2-2} > 1$.

Solution: The given equation

$$\frac{d^2y}{dt^2} + \frac{2}{\tau}\frac{dy}{dt} + \omega^2 y = 0.$$

represents the motion of a damped harmonic oscillator along the v-axis. Let us write it as

$$\frac{d^2y}{dt^2} = -\frac{2}{\tau}\frac{dy}{dt} - \omega^2 y.$$

The term d^2y/dt^2 is the instantaneous acceleration of the oscillator which is the sum of the acceleration due to damping force, $\frac{2}{\tau}\frac{dy}{dt}$, and the acceleration due to restoring force, $\omega^2 y$.

Since $\frac{2}{\tau}\frac{dy}{dt}$ is acceleration and $\frac{dy}{dt}$ is velocity, τ has the dimensions of time.

Let us write the solution of eq. (i) in the form

$$y = C\ e^{\alpha t}.$$

and

Then $$\frac{dy}{dt} = C\alpha\ e^{\alpha t}$$

and $$\frac{d^2y}{dt^2} = C\alpha^2\ e^{\alpha t}$$

Putting these values in eq. (i), we get

$$C\alpha^2\ e^{\alpha t} + \frac{2}{\tau}C\alpha\ e^{\alpha t} + \omega^2 C e^{\alpha t} = 0$$

or $$\alpha^2 + \frac{2}{\tau}\alpha + \omega^2 = 0$$

or $$\alpha = -\frac{1}{\tau} \pm \sqrt{\left(\frac{1}{\tau^2} - \omega^2\right)}$$

Hence the general solution of eq. (i) is

$$y = C_1 e^{\left[-\frac{1}{\tau} + \sqrt{\left(\frac{1}{\tau^2} - \omega^2\right)}\right]t}$$

$$+ C_2 e^{\left[-\frac{1}{\tau} - \sqrt{\left(\frac{1}{\tau^2} - \omega^2\right)}\right]t} \qquad \text{...(iii)}$$

When $\omega^2\tau^2 > 1$. the term $\sqrt{\left(\frac{1}{\tau^2} - \omega^2\right)}$ is imaginary. Then

$$\sqrt{\left(\frac{1}{\tau^2} - \omega^2\right)} = j\sqrt{\left(\omega^2 - \frac{1}{\tau^2}\right)} = j\omega'$$

where $\omega' = \sqrt{\left(\omega^2 - \frac{1}{\tau^2}\right)}$. Now. eq. (iii) becomes

$$y = C_1 e^{\left(-\frac{1}{\tau} + j\omega'\right)t} + C_2 e^{\left(-\frac{1}{\tau} - j\omega'\right)t}$$

$$= e^{-t/\tau}\left(C_1 e^{j\omega' t} + C_2 e^{-j\omega' t}\right)$$

$$= e^{-t/\tau}\{C_1(\cos\omega' t + j\sin\omega' t) + C_2(\cos\omega' t - j\sin\omega' t)\}$$

$$= e^{-t/\tau}\{(C_1+C_2)\cos\omega' t + j(C_1-C_2)\sin\omega' t)\}.$$

Let us put $C_1+C_2=a\sin\phi$ and $j(C_1-C_2)=a\cos\phi$ where a and ϕ are new constants. Then we have

$$y= e^{-t/\tau}(a\sin\phi\cos\omega' t + a\cos\phi\sin\omega' t)$$

or $$y=ae^{-t/\tau}\sin(\omega' t+\phi)$$

This is the solution of the given equation under the condition $\omega^2\tau^2 > 1$. This shows that the amplitude of motion, $ae^{-t/\tau}$ decreases with time exponentially, and the period is

$$T=\frac{2\pi}{\omega'}=\frac{2\pi}{\sqrt{\left(\omega^2-\frac{1}{\tau^2}\right)}}$$

The amplitude of motion at $t=0$ is a and at $t=\tau$ is a/e. That is, τ is the time-interval in which the amplitude decreases to $1/e$ of its initial value. Thus here τ is the mean lifetime of the (damped) oscillator.

3. *A particle of mass 40 gm experiences only a damping force proportional to its velocity. If its velocity is decreased from 100 cm/sec to 10 cm/sec in 23.0 second, calculate (i) relaxation time, (ii) the damping force when its velocity is 50 cm/see, (iii) the time in which its kinetic energy is reduced to one-tenth of its initial value, (iv) the total distance travelled if its initial velocity is 100 cm sec. (Given log. 10=2.30).*

Solution: Let v be the velocity of the particle at any instant t. Then the instantaneous damping force is

$$-bv$$

where b is a constant. If there is no other force acting, then by Newton's law (force=mass × acceleration), we have

$$m\frac{dv}{dt}=-bv$$

or $$\frac{dv}{dt} = -\frac{b}{m} v = -2rv,$$

where $b/m = 2\ r$. The expression may be written as

$$\frac{dv}{v} = -2r\ dt.$$

Integrating : $\log_e v = -2\ rt + A$ (constant).

If at $t=0$, we have $v = v_0$, then $A = \log_e v_0$. Then we have

$$\log_e v = -2\ rt + \log_e v_0$$

or $$\log_e \frac{v}{v_0} = -2\ rt$$

or $$\frac{v}{v_0} = e^{-2rt}$$

or $$\boxed{v = v_0\ e^{-2rt}.}$$

This is the expression for the (damped) velocity, which decreases exponentially with time.

Now, it is given that the velocity decreases from 100 cm/sec to 10 cm/sec in 23.0 seconds Hence, putting $\frac{v}{v_0} = \frac{10}{100} = \frac{1}{100}$ and $t=23.0$ second in eq. (i), we have

$$\frac{1}{100} = e^{-2r(23.0)}$$

or $$10 = e^{2r(23.0)}$$

or $$2\ r(23.0) = \log_e 10 = 2.30$$

$$\therefore\ 2\ r = \frac{2.30}{23.0} = \frac{1}{100}.$$

Eq. (i) therefore becomes

$$v = v_0 e^{-t/10}.$$

(i) Let τ be the relaxation time (the time interval in which the velocity reduces to i/e of its initial value). Thus substituting $\frac{v}{v_0} = \frac{1}{e}$ and $t=\tau$ in eq. (ii), we get

$$\frac{1}{e} = e^{-\tau/10}$$

or $$e^{-\tau} = e^{-\tau/10}$$

or $$\frac{\tau}{10} = 1$$

$$\tau = 10 \text{ sec.}$$

(ii) The damping force is

$$F_1 = -bv = -2\,rmv.$$

Here $2\,r = \frac{1}{100}$ sec^{-1}. m=40 gm. v=50 cm/sec.

$$\therefore\ F_d = -40 \times 50 = -200 \text{ dyne.}$$

(iii) The instantaneous kinetic energy is

$$K = \tfrac{1}{2} mv^2 = \tfrac{1}{2} mv_0^{\,2} e^{-t/5}.$$

If $K=K_0$ at t=0, then

$$K = K_0 e^{-t/5}.$$

Let t' be the time in which the kinetic energy is reduced to 1/10 of its initial value. Putting $\frac{K}{K_0} = \frac{1}{100}$ and t=t′ in the last expression, we get

$$\frac{1}{10} = e^{-t'/5}$$

or $$\frac{t'}{5} = \log_0 10 = 2.30$$

$$\therefore \quad t' = 5 \times 2.30\text{-}=11.5 \text{ sec.}$$

(iv) Putting $v=\frac{dx}{dt}$ in eq. (ii) and integrating, we get

$$x = v_0 e^{-t/10} (-10)+C \text{ (constant).}$$

At t=0, we have .x=0, so that C=$10v_0$.

$$\therefore \quad x = -10\, v_0\, e^{-t/10} + 10\, v_0 = 10\, v_0\, (i - e^{-t/10}).$$

At $t \to \infty$, $e^{-t/10} \to 0$, and $x \to 10\, v_0$.

Therefore, distance travelled if its initial velocity is 100 cm/sec, is

$$x = 10\, v_0 = 10 \times 100 = 1000 \text{ cm.}$$

4. A 2-gm particle is subjected to an elastic force of 30 dynes/cm and a frictional force of 5 dynes/(cm-sec⁻¹). If it is displaced through 2 cm and then released, find whether the resulting motion is oscillatory or not. If so, find its period.

Solution: The equation of motion of a particle of mass m subjected to an elastic force and a frictional force is

$$m\frac{d^2x}{dt^2} + b\frac{dx}{dt} + kx=0$$

where k is the elastic force per unit displacement and b the frictional force per unit velocity. Putting $\frac{b}{m} = 2$ r and $\frac{k}{m} = \omega^2$. we get

$$\frac{d^2x}{dt^2} + 2r\frac{dx}{dt}\, \omega^2 x = 0.$$

This represents an oscillatory motion provided $r < \omega$, and the time period is

$$T = \frac{2\pi}{\sqrt{(\omega^2 - r^2)}}$$

Here m=2 gm. k=30 dynes/cm and b=5 dynes.(cm-sce^{-1}

$$r = \frac{b}{2m} = \frac{5}{2 \times 2} = 1.25$$

and $$\omega = \sqrt{\left(\frac{k}{m}\right)} = \sqrt{\left(\frac{30}{2}\right)} = \sqrt{(15)} = 3.9$$

Thus $r < \omega$. Hence the motion is oscillatory.

The time-period is

$$T = \frac{2\pi}{\sqrt{(\omega^2 - r^2)}} = \frac{2 \times 3.14}{\sqrt{(15 - 1.5625)}}$$

$$= \frac{2 \times 3.14}{3.67} = 1.71 \text{ sce.}$$

5. *A body of mass 0.2 kg is hung from a spring of constant 80 nt/m. The body is subject to a resistive force given by bv, where v is the velocity in m/s. Calculate the value of the undamped frequency, and the value of* τ. *if the damped frequency is* $\sqrt{3/2}$ *of the undamped frequency.*

Solution: The undamped frequency of a mass m suspended by a spring of force-constant k is given by

$$n = \frac{1}{2\pi}\sqrt{\left(\frac{k}{m}\right)}$$

Here $m = 0.2\ kg$ and $k = 80$ nt/m.

$$\therefore \quad n = \frac{1}{2\pi}\sqrt{\left(\frac{80}{0.2}\right)} = \frac{10}{2\pi} = 3.18 \text{ sec}^{-1}$$

The damped frequency n′ (say) is given by

$$n' = \frac{1}{2\pi}\sqrt{(\omega^2 - r^2)},$$

where $\omega = \sqrt{(k/m)}$ and r—$b/2m$.

$$\therefore \quad n' = \frac{1}{2\pi}\sqrt{\left(\frac{k}{m} - r^2\right)}$$

But $n = \left(\sqrt{3/2}\right)$

$$\therefore \quad n' = \frac{1}{2\pi}\sqrt{\left(\frac{k}{m} - r^2\right)} = \frac{\sqrt{3}}{2}\frac{10}{2\pi}$$

or $$\frac{k}{m} - r^2 = 300$$

Here $\frac{k}{m} = \frac{80}{0.2} = 400.$

$$\therefore \quad 400 - r^2 = 300$$

or $r = 10.$

Therefore, the relaxation time is

$$\tau = \frac{1}{2r} = \frac{1}{20} = 0\text{-}05 \text{ sec.}$$

6. If the relaxation time of a damped harmonic oscillator is 50 second, find the time in which (i) the amplitude falls to 1/e time the initial value, (ii) energy of the system falls to 1/e times the initial value, (iii) energy falls to 1/e⁴ of the initial value.

Solution: The equation of a damped harmonic oscillator is

$$x = ae^{-rt}\sin(\omega' t + \phi),$$

where r is a damping constant. The relaxation time is

$$\tau = \frac{1}{2r} = 50 \text{ second}$$

so that

$$r = \frac{1}{100} \text{ sec}^{-1}.$$

(i) The damped amplitude at any instant is *aer"*. Let a_0 be the amplitude at t=0, and *aje* at *t*. Then we can write

$$a_0 = a$$

and $$\frac{a_0}{e} = ae^{-rt} = a_0 e^{-rt}.$$

From this, we have

$$e^{-1} = e^{-rt}$$

$$\therefore \quad t = \frac{1}{r} = 100 \text{ second.}$$

(ii) The energy (or velocity) falls to 1/e times the initial value in the relaxation time. Thus in this case

$$t = \tau = 50 \text{ second.}$$

(iii) The energy of a damped harmonic oscillator is given by

$$E = E_0 e^{-2rt},$$

where E_o is the energy at $t = 0$. Let t′ be the time during which the energy falls to E/e^4. Then we have

$$\frac{E_0}{e^4} = E_0 e^{-2rt'}$$

or $$e^{-4} = e^{-2rt'}$$

or $$4 = 2rt'.$$

$$\therefore \quad t' = \frac{2}{r} = 2 \times 100 = 200 \text{ second.}$$

7. The amplitude of an oscillator of frequency 200 cycles/sec falls to 1/10 of its initial value after 2000 cycles. Calculate (i) its relaxation time, (ii) its quality factor, (iii) time in which its energy falls to 1/10 of its initial value, (iv) damping constant.

Solution: The oscillator having a frequency of 200 cycle/sec, completes 2000 cycles in 10 sec. The instantaneous amplitude of the (damped) oscillator is ae^{-rt}, where r is a damping constant. If a_0 be its initial amplitude (at t=0) and $a_0/10$ after 10 sec, then we have

$$a_0 = a$$

and $$\frac{a_0}{10} = ae^{-10r} = a_0 e^{-10r}.$$

From this we have

$$10 = e^{10r}.$$

Taking logarithm:

$$\log_e 10 = 10r$$

or $$10r = 2.3 \log_{10} 10 = 2.3$$

$$\therefore \quad r = \frac{2.3}{10} = 0.233$$

(i) No, the relaxation time is

$$\tau = \frac{1}{2r} = \frac{1}{1 \times 0.23} = 5.0 \text{ sec.}$$

(ii) The quality factor is

$$Q = \omega\tau = 2\pi n\tau$$

$$= 2x3.14x200x2.174 = 2730.$$

(iii) The energy of the oscillator is

$$E = E_0 e^{-2rt}$$

Let t' be the time after which the energy falls to $E_0/10$. Then

$$\frac{E_0}{10} = E_0 e^{-2rt}$$

or $$10 = e^{2rt'}$$

or $$2rt' = \log_4 10 = 2.3$$

or $$t = \frac{2.3}{2r} = \frac{2.3}{2 \times 0.23} = 5.0 \text{ sec.}$$

(iii) The damping constant

$$r = 0.23 \text{ (as determined above).}$$

8. The quality factor Q of a tuning fork is 5.0 × 10⁴. Calculate the time-interval after which its energy becomes 1/10 of its initial value. The frequency of the fork is 300 sec⁻¹ (log. 10=2.3).

Solution: The quality factor of a damped oscillator is given by

$$Q = \omega\tau,$$

where ω is angular frequency and τ is relaxation time. Therefore

$$\tau = \frac{Q}{\omega}.$$

Here $Q=5.0 \times 10^4$ and $\omega = 2\pi n = 600\,\pi$ sec^{-1}.

$$\therefore \quad \tau = \frac{5.0\times10^4}{600\pi}\text{ sec.}$$

Now, the energy of the damped system is given by

$$E = E_0 e^{-t/\tau} \qquad \left[\because \ \tau = \frac{1}{2r}\right]$$

Let t′ be the time after which the energy becomes $\frac{1}{10}$ of the initial value. Then putting $E/E_0 = \frac{1}{10}$ and t=t′ in the last expression, we get

$$\frac{1}{10} = e^{-t'/\tau}$$

or

$$10 = e^{t'/\tau}$$

or

$$\log_e 10 = \frac{t'}{\tau}$$

$$\therefore \quad t' = \tau \log_e 10$$

$$= \frac{5.0\times10^4}{600\times3.14}\times2.3 = 61 \text{ sec.}$$

9. Q of a sonometer wire is 2 × 10. On plucking, it executes 240 vibrations per second. Calculate the time in which the amplitude decreases to $1/e^2$ of the initial value.

Solution: The quality factor is

$$Q = \omega\tau$$

Here $Q=2\times10^3$ and $\omega = 2\pi n = 2 \times 3.14 \times 240 \text{ sec}^{-1}$.

$$\therefore \quad \tau = \frac{Q}{\omega} = \frac{2\times10^3}{2\times3.14\times240} = 1.327\text{sec.}$$

Now, if a_0 be the initial amplitude (at t=0) and a_0/e^2 the amplitude after time t, we have

$$\frac{a_0}{e^2} = a_0 e^{-rt}$$

or $$e^{-2} = e^{-rt}$$

or $$2 = rt$$

or $$t = \frac{2}{r} = 4\tau \qquad \left[\because \; \tau = \frac{1}{2r}\right]$$

$$= 4 \times 1.327 = 5.3 \text{ sec.}$$

10. The oscillations of a tuning fork of frequency 200 cps die away to 1/e times their amplitude in 1 second. Show that the reduction in frequency due to air damping is exceedingly small.

Solution: The damped amplitude at any time t is ae^{-rt}. where a is a constant. If a_0 be the amplitude at t=0 and a_0/e at time t, then we have

$$\frac{a_0}{e} = a_0 e^{-rt}.$$

This gives $$t = \frac{1}{r}.$$

Here t=1 second.

$$\therefore \quad r = 1 \text{ sec}^{-1}$$

The angular frequency in presence of damping is

$$\sqrt{(\omega^2 - r^2)} = \sqrt{(\omega^2 - 1)}.$$

Now $\omega = 2\pi n$ = 2 × 3.14 × 200= 1256, so that $\omega^2 \approx 1.6 \times 10^6$.

$$\therefore \quad \text{damped frequency} = \sqrt{(1.6 \times 10^6 - 1)}$$

$$\simeq \sqrt{(1.6 \times 10^6)} \cong \sqrt{\omega^2} \simeq \omega,$$

which is almost same as undamped frequency. Thus the effect of damping on frequency is exceedingly small.

11. A damped vibrating system starting from rest has an initial amplitude of 20 cm which reduces to 2 cm after 100 complete oscillations each of period 2.3 sec. Find the logarithmic decrement of the system.

Solution: The logarithmic decrement λ is the logarithm of the ratio of two amplitudes of oscillations which are separated by one period. Here the amplitude ratio of oscillations separated by 100 periods is $\frac{20}{2}$. Thus

$$\lambda = \frac{1}{100} \log_e \frac{20}{2}$$

$$= \frac{1}{100} \log_e 10$$

$$= \frac{1}{100} \times 2.3 = 0.023.$$

10

Two-Body Oscillations

Two Body Harmonic Oscillator

Let us consider two masses m_1 and m_2 connected by a massless spring of force constant k. If the masses are slightly displaced from equilibrium in opposite directions, the spring extends (or contracts) and exerts a linear restoring (elastic) force on both the masses.

Therefore, the masses, when released, oscillate about their equilibrium positions, their centre of mass remaining at rest. Such a system is called a 'two-body oscillator'

A diatomic molecule is a stable system of two nuclei (masses) connected by a spring-like force. Thus it is an example of a two-body harmonic oscillator. Let x_1 and x_2 be the distances of m_1 and m_2 respectively from the origin O of an inertial frame of reference at an instant during the oscillation. Let l be the equilibrium length of the spring and x the instantaneous change in length. Then, from the figure, we have

$$x_2 - x_1 = l + x. \qquad \text{...(i)}$$

If x is positive, the spring is stretched, and if x is negative, the spring is compressed. Suppose in the figure, x is positive.

Now, the restoring force exerted by the spring on each mass a magnitude kx. If x is positive, the mass m_l is displaced to the left from its equilibrium position and the mass m_2 to the right. That is, the spring is tending *to* pull m_1 to the right (in positive direction) and m_2 to the left (in negative direction). Thus, the (restoring) force on m_1 is $F_1=kx$, and on m_2 is $F_2=-kx$. Hence, from Newton's second law (force=mass X acceleration), the equations of motion are

$$F_1 = kx = m_1 \frac{d^2 x_1}{dt^2}$$

and
$$F_2 = kx = m_2 \frac{d^2 x_2}{dt^2}$$

Let us multiply the first equation by m_2 and the second equation by m_1 and then subtract the first from the second. The result is

$$m_1 m_2 \left(\frac{d^2 x_2}{dt^2} - \frac{d^2 x_1}{dt^2} \right) = -\text{k}(m_1+m_2)\, x$$

or
$$\frac{d^2}{dt^2}(x_2-x_1) = -k\, \frac{m_1+m_2}{m_1 m_2}\, x.$$

But $x_2-x_1=x+l$ from eq. (i).

$$\therefore \quad \frac{d^2}{dt^2}(l+x) = -k\, \frac{m_1+m_2}{m_1 m_2}\, x.$$

Because l is a constant, $\frac{d^2 l}{dt^2}$ is zero. Therefore the last equation reduces to

$$\frac{d^2 x_1}{dt^2} = -k \frac{m_1 + m_2}{m_1 m_2} x \qquad \text{...(ii)}$$

The quantity $\frac{m_1 m_2}{m_1 + m_2}$ is defined as the 'reduced mass' of the system and is denoted by μ. That is

$$\mu = \frac{m_1 m_2}{m_1 + m_2}$$

The eq. (ii) may now be written as follows :

$$\frac{d^2 x}{dt^2} = -\frac{k}{\mu} x$$

or

$$\frac{d^2 x}{dt^2} = -\omega^2 x, \qquad \text{...(iii)}$$

where $\omega^3 = k/\mu$. This equation has exactly the same form as of a single-body simple harmonic oscillator, with the difference that here x is the *relative* displacement of the two bodies from their equilibrium separation. Hence the *relative* motion of two masses m_1 and m_2 is just the same as the motion of a simple harmonic oscillator of mass $\mu = \frac{m_1 m_2}{m_1 + m_2}$.

The solution of the differential eq. (iii) is of the form

$$x = a \sin(\omega t + \phi)$$

where a and ϕ are constants. The time-period and frequency are

$$T = \frac{2\pi}{\omega} = 2\pi\sqrt{\left(\frac{\mu}{k}\right)}$$

and

$$n = \frac{1}{T} = \frac{1}{2\pi}\sqrt{\left(\frac{k}{\mu}\right)}.$$

Thus the frequency of a two-body oscillator is exactly the same as the frequency of a single-body oscillator of mass μ connected with one end of the same spring, the other end of which is fixed in an inertial frame of reference. In other words, in two-body oscillation, one body moves relative to the other as if the other body were fixed and the mass of the moving body were reduced to μ.

Vibrations of a Diatomic Molecule

A diatomic molecule is a system of two point-masses (nuclei) connected through a spring like force. If the distance *r* between the nuclei changes, the potential energy *U* of the molecule changes. If at infinite distance between the nuclei, the potential energy is considered zero, a graph between energy *U* and distance *r* is as shown in figure given below. The energy is minimum, equal to U_0, in the equilibrium state to the molecule when the distance between the nuclei is r_0. The curve is almost parabolic near the bottom.

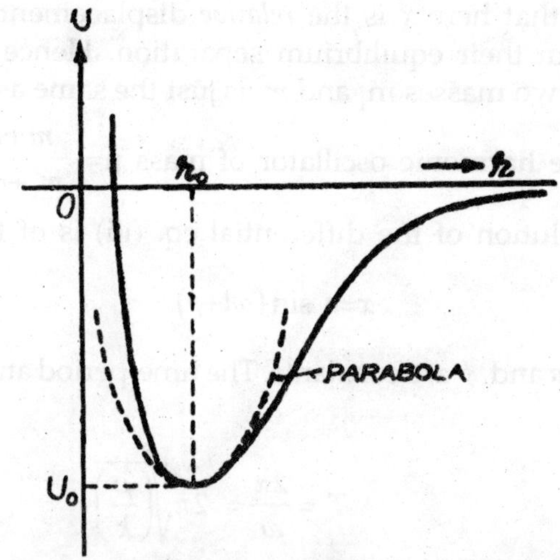

Hence, in this region, the equation of the curve may be written as

$$U = -U_0 + \tfrac{1}{2} k (r-r_0)^2,$$

where r is the separation of nuclei at any instant, and k is a constant. The force between the nuclei is therefore

$$F = -\frac{dU}{dr}$$

$$= -k\,(r - r_0).$$

The force is thus proportional to $(r-r_0)$, the displacement of the separation between the nuclei from the equilibrium separation. Hence the nuclei perform simple harmonic oscillations along the internuclear axis, and the molecule may be treated as a two-body harmonic oscillator, with a natural frequency given by

$$v = \frac{1}{2\pi}\sqrt{\frac{k}{\mu}},$$

where k is the force-constant and μ is the reduced mass of the molecule.

Vibrational Energy States of the Molecule: The vibrations of the nuclei of a diatomic molecule, like HCl, are in fact responsible for the spectrum of the molecule in the near infrared region.

Classically, an oscillator has an energy of vibration which increases continuously as the amplitude of vibration increases. But quantum mechanically, an oscillating molecule can possess only certain discrete values of energy and no others. These permissible values of vibrational energy are given by

$$E = (n + \tfrac{1}{2})\, h\nu,$$

where n is a positive integer (n=0, 1, 2, 3,.......) and is called the 'vibrational quantum number', h is Planck's constant and ν is the natural frequency of the molecule.

Putting n=0, 1, 2,..... in the above expression, we get

$$E = \tfrac{1}{2}h\nu,\ \tfrac{3}{2}h\nu,\ \tfrac{5}{2}h\nu,\ \ldots$$

These are the values of vibrational energy which the molecule can possess. In other words, the molecule can occur in discrete vibrational energy states. These states can be represented by a set of parallel equispaced straight lines having a common separation hν, in the potential energy curve. For high values of *n*, however, the potential curve departs from the parabolic form and the lines gets closer. These lines are called 'energy level' and the diagram is called an 'energy level diagram'.

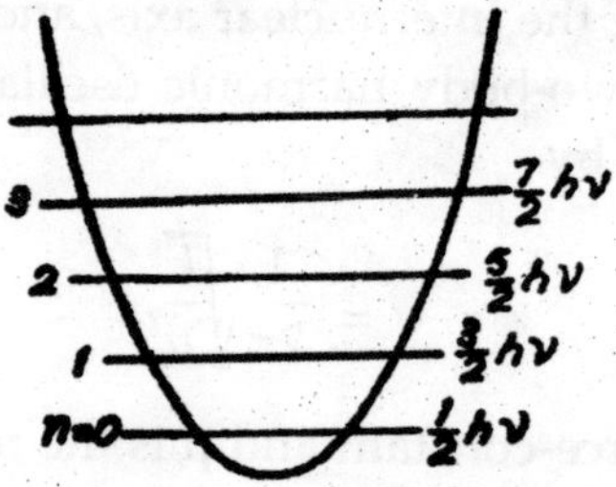

The molecule can pass from one energy state to the other such that *n* changes by 1 (selection rule). In passing from a lower to a higher state it absorbs energy $h\nu$, while in passing from a higher to a lower state it emits energy $h\nu$ in the form of electro-magnetic radiation.

Problems

1. Two bodies of masses 1.0 kg and 3.0 kg are joined by a massless spring of force constant *k*=25 newton/metre. The bodies are pulled apart and released. Compute the frequency of the resulting oscillations.

Solution: The reduced mass of the oscillating system is

$$\mu = \frac{m_1 m_2}{m_1+m_2} = \frac{(1.0)(3.0)}{1.0+3.0} = 0.75 \text{ kg}.$$

The frequency of the oscillator is

$$n = \frac{1}{2\pi}\sqrt{\left(\frac{k}{\mu}\right)}$$

$$= \frac{1}{2 \times 3.14}\sqrt{\left(\frac{250}{0.75}\right)} = 2.9 \text{ sec}^{-1}$$

2. Calculate the fundamental frequency of vibration of H^1Cl^{35} molecule. The force-constant of HCl is 5.0×10^5 dyne/cm.

Solution: If m_H and m_{cl} be the masses of the H^1 and Cl^{35} atoms, then the reduced mass of the molecule is

$$\mu_{HCl} = \frac{m_H m_{cl}}{m_H + m_{cl}}.$$

The atomic weight of H is 1, so that the mass of one H atom is $\frac{1}{6.023 \times 10^{23}}$ gm, where 6.023×10^{23} is Avogadro's number.

Similarly, the mass of one Cl atom is $\frac{35}{6.023 \times 10^{23}}$ gm. Thus

$$\mu HCl = \frac{\frac{1}{6.023 \times 10^{23}} \times \frac{35}{6.023 \times 10^{23}}}{\frac{1}{6.023 \times 10^{23}} + \frac{35}{6.023 \times 10^{23}}}$$

If k be the force-constant, then the frequency of the molecule is given by

$$n = \frac{1}{2\pi}\sqrt{\left(\frac{k}{\mu}\right)}$$

$$= \frac{1}{2 \times 3.14}\sqrt{\left(\frac{5.0 \times 10^5 \text{ dyne/cm}}{1.6 \times 10^{-24} \text{ gm}}\right)}$$

$$= \frac{1}{2 \times 3.14}\sqrt{\left(\frac{500}{16}\right)} \times 10^{14}$$

$$= 0.89 \times 10^{14} = 8.9 \times 10^{13} \text{ sec}^{-1}.$$

3. From spectroscopic measurements the fundamental vibrational angular frequency of the HF molecule is 7.55×10^{14}

rad/sec. Find (i) force-constant for HF molecule, (ii) work done is stretching the atoms apart by 0.5 Å- Given mH=1.66 $\times 10^{-24}$ gm, m_F—19 × 1.66 × 10^{-24} gm and equilibrium internuclear distance r_o=1 Å.

Solution: The reduced mass of the *HF* molecule is

$$\mu = \frac{m_H m_F}{m_H + m_F} = \frac{(1.66\times10^{-24})(19\times1.66\times10^{-24})}{(1.66\times10^{-24})+(19\times1.66\times10^{-24})}$$

$$= \frac{19}{20}\times(1.66\times10^{-24}) = 1.58\times10^{-24}\text{gm.}$$

(i) Let *k* be the force-constant. Its angular frequency is

$$\omega = 2\pi n = \sqrt{\left(\frac{k}{\mu}\right)}$$

$$\therefore\; k = \mu\omega^2.$$

Here μ = 1.58 × 10^{-24} gm and ω = 7.55 × 10^{14} rad/sec.

$\therefore$ k = (1.58 × 10^{-24}) × (7.55 × $10^{14})^2$=9.0 × 10^5 dyne/cm.

(ii) The work done in stretching the atoms through a distance *x* is given by

$$W = \tfrac{1}{2}kx^2.$$

Here μ = 9.0 × 10^5 dyne/cm and x=0.5 *k Å*=0.5 × 10^{-8}cm.

$$\therefore\; W\tfrac{1}{2}\times(9.0\times10^5)\times(0\text{-}5\times10^{-8})^2$$

$$=1.125\times10^{-11}\text{ erg.}$$

4. The vibrational energy levels in HCl molecule made from H and Cl isotopes of atomic weights 1 and 35 are separated by 0.36 eV. Obtain (i) frequency and wavelength emitted in a pure vibrational transition, (ii) force-constant, (iii) separation of energy levels if HCl is made from D^2 and Cl^{37} isotopes. Mass of hydrogen atom is 1-6 × 10^{-24} gm.

Solution: **(i)** By quantum theory, the frequency of radiation emitted in a vibrational transition is given by

$$\Delta E = h\nu$$

where ΔE is the energy difference between vibrational levels.

Here $\Delta E = 0.36$ eV $= 0.36 \times (1.6 \times 10^{-12})$ erg; Planck's constant h is 6.6×10^{-27} erg-sec.

$$\therefore \quad \nu = \frac{\Delta E}{h} = \frac{0.36 \times (1.6 \times 10^{-12})\text{erg}}{6.6 \times 10^{-27}\text{ erg-sec}} = 8.7 \times 10^{13}\text{sec}^{-1}$$

The corresponding wavelength is

$$\lambda = \frac{c}{\nu} = \frac{3 \times 10^{10}\text{ cm sec}^{-1}}{8.7 \times 10^{13}\text{ sec}^{-1}} = 3.4 \times 14^{-4}\text{ cm} = 3.4 \times 10^{4}\text{ Å}.$$

(ii) The vibrational frequency is

$$\nu = \frac{1}{2\pi}\sqrt{\left(\frac{k}{\mu}\right)}$$

$$\text{Here } \mu = \frac{m_1 m_2}{m_1 + m_2} = \frac{1 \times 35}{1 + 35} \times \text{mass of } H \text{ atom}$$

$$\therefore \quad k = 4\pi^2 \nu^2 \mu^2$$

$$= 4 \times (3.14)^2 \times (8.7 \times 10^{13})^2 \times (1.56 \times 10^{-24})$$

$$= 4.6 \times 10^{5}\text{ dyne/cm}.$$

(iii) Let μ and μ' be the reduced masses of H^1C1^{35} and D^2Cl^{37} respectively. Then

$$\frac{\mu}{\mu'} = \frac{(1 \times 35)/(1 + 35)}{(2 \times 37)/(2 + 37)}$$

$$\frac{35}{36} \times \frac{39}{74} = 0.51.$$

If ν and ν' be the corresponding frequencies, then

$$\frac{\nu'}{\nu} = \sqrt{\left(\frac{\mu}{\mu'}\right)} = \sqrt{(0.51)} = 0.72$$

Finally, if ΔE and $\Delta E'$ be the corresponding energy separations, then we have

$$\frac{\Delta E'}{\Delta E} = \frac{\nu'}{\nu} = 0.72$$

$$\therefore \Delta E' = 0.72 \; \Delta E' = 0.72 \times 0.36 = 0.26 \text{ eV.}$$

Bibliography

Anderson, J.R. Stevens D.E. and Julian P.R.: *Temporal Variations of the Tropical 40-50 Day Oscillation:* Monthly Weather Review,1984.

Bantzer, C.H. and Wallace J.M.: *Intraseasonal Variability in Tropical Mean Temperature and Precipitation,* Journal of The Atmospheric Sciences, 1996.

Blade, I. and Hartmann D.L.: *Tropical Intraseasonal Oscillations in a Simple Nonlinear Model,* Journal of the Atmospheric Sciences,1993.

Blade, Ileana: *Dynamics, Thermodynamics and Extratropical Interactions of Tropical Intraseasonal Oscillations,* University of Washington, Washington, 1994.

Chang, F.C. and Lau K.M.: *Tropical Intraseasonal Oscillation and its Prediction by the NMC Operational Model,* Journal of Climate, 1992.

Djurovic, D. and Paquet P.: *New Indications of Oscillation in Solar Activity, Atmospheric Circulation and Earth's Rotation,* Solar Physics, 1995.

Eischeid, J. Weickmann, K.M. and Khalsa S.J.S.: *The Atmospheric Angular Momentum Cycle During the Tropical Madden Julian Oscillation,* Monthly Weather Review, 1992.

Ferreira, R.N Schubert W.: *Dynamical Aspects of Twin Tropical Cyclones Associated with the Madden Julian Oscillation,* Journal of the Atmospheric Sciences, 1996.

Gao, X.H. and Stanford, J.L.: *Low Frequency Oscillations in Total Ozone Measurements,* Journal of Geophysical Research, 1990.

Gelaro, R. and Murphree T.: *Observed and Simulated Northern Hemisphere Intraseasonal Circulation Anomalies and the Influence of Model Bias,* Monthly Weather Review, 1996.

Gutzler, D.S. and Ponte, R.M.: *Oscillations in the Western Tropical Pacific,* Geophysical Research Letters, 1992.

Hartmann, D.L.; Michelsen M.L. and Klein S.A.: *Seasonal Variations of Tropical Intraseasonal Oscillations,* Journal of the Atmospheric Sciences, 1992.

Hartten, Leslie Marie: *Westerly Wind Bursts,* University of Wisconsin, Madison, 1993.

Held, I.M. Hemler, R.S. and Ramaswamy V.: *Radiative Convective Equilibrium with Explicit two Dimensional Moist Convection,* Journal of the Atmospheric Sciences, 1993.

Kessler, W.S. and McPhaden M.J.: *Forcing of Intraseasonal Kelvin Waves in the Equatorial Pacific,* Journal of Geophysical Research, 1995.

Kiladis, G.N. and Wheeler M.: *Horizontal and Vertical Structure of Observed Tropospheric Equatorial Rossby Waves,* Journal of Geophysical Research, 1995.

Kirtman, B. and Vernekar, A.: *On Wave CISK and the Evaporation Wind Feedback for the Madden Julian Oscillation,* Journal of the Atmospheric Sciences, 1993.

Lau, K.M. Sheu P.J. Schubert S. and Ledvina D.: *Evolution of Large Scale Circulation During Toga Coare Model Intercomparison and Basic Features,* Journal of Climate, 1996.

Liebmann, B.: *Organization of Convection within the Madden Julian Oscillation,* Journal of Geophysical Research, 1994.

Magana, V.: *Oscillations in Atmospheric Angular Momentum at Various Latitudes,* Journal of Geophysical Research, 1993.

Marcus, S.L and Ghil M.: *The Extratropical 40-Day Oscillation in the Ucla General Circulation Model*, Journal of The Atmospheric Sciences, 1996.

Meehl, G.A. Kiladis G.N. and Weickmann K.M.: *Modulation of Equatorial Subseasonal Convective Episodes by Tropical Extratropical Interaction in the Indian and Pacific Ocean Regions*, Journal of Geophysical Research Atmospheres, 1996.

Mertz, Gordon James: *Atmospheric and Oceanic 40 To 50 Day Oscillations in the Source Region of the Somali Current*, University of British Columbia, Canada, 1985.

Mysak, L.A. and Mertz, G.J.: *Oscillation in the Source Region of the Somali Current During 1976*, Journal of Geophysical Research, 1984.

Njau, E.C: *Generalized Theory of Enso and Related Atmospheric Phenomena*, Renewable Energy,1996.

Paquet, P. and Djurovic D.: *The Frequency Composition of the Madden Julian Oscillation*, Solar Physics, 1985.

Park, C.K. and Schubert S.D.: *Remotely Forced Intraseasonal Oscillations over the Tropical Atlantic*, Journal of the Atmospheric Sciences, 1993.

Ponte, R.M. and Gutzler D.S.: *The Madden Julian Oscillation and the Angular Momentum Balance in a Barotropic Ocean Model*, Journal of Geophysical Research, 1991.

Rasmusson, E.M. and Arkin P.A.: *A Global View of Large Scale Precipitation Variability*, Journal of Climate,1993.

Ready, S. and Woodcock, F.: *The South Pacific and Southeast Indian Ocean Tropical Cyclone Season 1989-1990*, Australian Meteorological Magazine, Australia, 1992.

Reynolds, C. and Murakami T.: *Tropical and Midlatitude 45-Day Perturbations over the Western Pacific During the Northern Summer*, Journal of the Meteorological Society of Japan, Japan, 1996.

Sheng, J.: *The Madden Julian Oscillation in the Canadian Climate Centre General Circulation Model*, Climate Dynamics, 1995.

Suppiah, R.: *ENSO Phenomenon and 30-50 Day Variability in the Australian Summer Monsoon Rainfall,* International Journal of Climatology, 1993.

Swinbank, R. Palmer T.N. and Davey M.K.: *Numerical Simulations of the Madden and Julian Oscillation,* Journal of the Atmospheric Sciences, 1989.

Takayabu, Y.N and Lau K.M and Sui C.H.: *Observation of a Quasi-2-Day Wave During Toga Coare,* Monthly Weather Review, 1999.

Thapliyal, V. Kripalani R.H. and Singh S.V.: *Global structure of the Madden Julian Oscillations During Two Recent Contrasting Summer Monsoon Seasons Over India,* Meteorology and Atmospheric Physics, 1993.

Tsing, Chang Chen and Ming Cheng Yen: *A Study of the Diabatic Heating Associated with the Madden Julian Oscillation,* Journal of Geophysical Research, 1991.

Vergara, J.A. and Fuenzalida H.A.: *Intraseasonal Oscillations in the Subtropical Subsidence of the Eastern Pacific,* Contributions to Atmospheric Physics, 1993.

Vincent, R.A.: *First Observations of Intraseasonal Oscillations in the Equatorial Mesosphere and Lower Thermosphere,* Geophysical Research Letters, 994.

Weickmann, K.M.: *El Nino/Southern Oscillation and Madden Julian Oscillations During 1981-1982,* Journal of Geophysical Research, USA, 1991.

Xiao, Ziniu and Li Chongyin: *Numerical Simulation of the Atmospheric low Frequency Eleresponse to External Forcing,* Scientia Atmospherica Sinica,1993.

Zhang, C.D and Hendon H.H.: *Propagating and Standing Components of the Intraseasonal Oscillation in Tropical Convection:* Journal of the Atmospheric Sciences, 1996.

Zhang, Chidong: *Atmospheric Intraseasonal Variability at the Surface in the Tropical Western Pacific Ocean,* Journal of the Atmospheric Sciences, 1996.

Index

P

Q

R

S

T

V

W

❑❑❑